CAR AND DRIVER

Sports Cars Illustrated changed its name to Car and Driver in April 1961

CAR AND DRIVER

Car and Driver is a living, breathing entity. Although some of our older editions may be a bit cranky and occasionally stiff-legged, they do live on for an eternity (if only on our shelves). We do love to share our previous work, but it's obviously not possible to crank up the printing presses on a whim. Instead, we've produced this series of books, each of which encompasses virtually everything said about a particular subject during a given period in Car and Driver.

We hope you enjoy these collections. They have not been edited or updated in any way, so this is vintage Car and Driver at its finest (we think).

Printed in Hong Kong

SCI ROAD TEST:

Chevrolet Corvette

Here the author takes a long, fast turn at 60 mph. Note that the rear dips more than front, giving car an understeer effect. Stiffer shocks and springs on the back end would add stability on bends.

By KARL LUDVIGSEN

CHANCES are that by the time you read this the '56 Corvette will have made a profound impression on the whole sports car world, and after having had one under me for a couple of days I will be the last to be surprised. This very early production model showed a willingness and ability to be driven fast and hard under almost all conditions and demonstrated an even greater potential for competitive use. In my opinion, the Corvette as it stands is fully as much a dual-purpose machine as the

Photos by Don Typond

The Halibrand type knock-offs look like the McCoy, but are actually wheel discs. The real thing is being planned as possible optional equipment.

Cruising at 85, the Corvette produces wind noise over top. However, at such speeds there's bound to be wind disturbance on any car.

Although automatically raised and lowered, the top's back window has to be pulled down by hand. This is no problem since the well fitted top clamps down quickly sans fuss.

stock Jaguar, Triumph, or Austin-Healey. Without qualification, General Motors is now building a sports car.

Unfortunately, at this writing accurate information both on the Corvette itself and on their future plans is not available, and the air is rife with rumor. SCI's test car was chassis #1002, and was obtained from the Chevrolet Motor Division through the combined efforts of Shelly Spindel and Alvin Schwartz Chevrolet of Brooklyn, N. Y. Finished in two-tone turquoise with a matching hard top and a white convertible top, it was a real traffic-stopper, and was specifically destined to make a New York TV appearance. As such, it had the full range of options, including whitewalls, the hard top, power windows, radio, heater and windshield washer. To our joy it had the close-ratio stick shift, but less happily had the higher, 3.27:1 rear end ratio. It was, all in all, a lot of car and I regret that at this time Chevrolet was not ready to discuss prices. There is little question, though, that it is to be competitive with the Thunderbird.

Now that the "dual-purpose" claim has been made, it should be backed up. Those accustomed to GM products may tend to take the creature comforts for granted, but any owner of an older Corvette will readily testify that those cars could be uninhabitable at times. Much effort has been expended to rectify this, and it has paid off in full.

Entry and exit over the wide sill on the passenger side is easy, for a sports car, but as the driver slides under the steering wheel he becomes aware of one of the car's few major faults. While it is handsome, and provided with more than enough finger ribbing, the wheel is too close to the driver and is non-adjustable. Ex-Jag drivers may find the position natural, but I personally felt that more arm room would be useful, particularly for competition. You also sit close enough to the door for the integral arm rest to be in the way.

The seats themselves are very handsome, and very deceptive. They look like a true bucket type, and the seat bottoms are comfortable enough, but the backs are bolt upright and provide no lateral support for the torso. Adjustment of rake and a more definite "bucket" would improve them greatly and would obviate a certain amount of fatigue that now occurs. Fore-and-aft adjustment is not extensive, there being just enough room for a six-footer. Leg room is excellent; the left foot can roam about under the suspended pedals, and the brake and throttle are well-placed for heel-and-toe downshifting.

Headroom is also at the bare minimum for six feet of height, with slightly more room under the soft top. In general, the Corvette has very little interior room for such a large car, and it seems that neither GM nor Ford have yet completely solved the sports car seating problem.

Driving the Corvette with the top down is very pleasant, the windshield giving good protection to the top and side. You sit high enough to rest your elbow comfortably on the door, if so inclined. The power windows are handy and reliable, but the power-operated top qualifies as the most fascinating mechanism I have seen on any car. The lid rises, the top emerges, and the lid closes again as the control button is pressed. You must then pull down and clamp the back window section by hand, and close two front latches. It's all very easy, and the finished product is attractive and tight. It can be stowed away just as simply.

The hard top is also easy to install, having two clamps at the front and three bolts and two locating dowels at the rear. It is well finished and padded, and provides unobstructed vision. Light and easy to transport, the top's main fault is arbitrary sealing at the sides of the rear deck.

Instrument panel layout is identical to last year's cars, and has many of the same faults. All the secondary instruments, including the tachometer, are very difficult to read

CAR AND DRIVER

ON CORVETTE

1956–1967

Reprinted From
Car and Driver Magazine

ISBN 0 946489 90 4

Published By
Brooklands Books with permission of Car and Driver

CAR AND DRIVER

Titles in this series

CAR AND DRIVER ON CORVETTE 1956-1967
CAR AND DRIVER ON CORVETTE 1968-1977
CAR AND DRIVER ON CORVETTE 1978-1982
CAR AND DRIVER ON FERRARI 1955-1962
CAR AND DRIVER ON FERRARI 1963-1975
CAR AND DRIVER ON FERRARI 1976-1983

Titles in preparation will cover: Dodge, Jaguar, Lamborghini, Mustang, Plymouth, Pontiac, Porsche etc.

Distributed By

Car and Driver
3460 Wilshire Blvd,
Los Angeles,
California 90010

Brooklands Book Distribution Ltd.
Holmerise, Seven Hills Road,
Cobham, Surrey KT11 1ES,
England

Twin four-barrel carburetors squat on power plant ready to help deliver 225 bhp at 5200 rpm. At low speeds, only rear carburetor functions. BELOW—Hood raises from rear, reducing possible lifting at high speed. Engine compartment is more accessible for shop work.

A lack of bumpers at the rear will make any Corvette driver over-cautious when backing. Exhaust tubes are in for abuse.

SPECIFICATIONS
CHEVROLET CORVETTE

ENGINE

Cylinders	V8
Bore and stroke	3.75 in x 3.00 in (95 mm x 76 mm)
Displacement	265 cu in (4340 cc)
Compression ration	9.25:1
Max. horsepower	225 bhp @ 5200 rpm
Max. torque	270 lb ft @ 3600 rpm
Max. b.m.e.p.	154 psi

CHASSIS

Wheelbase	102 in
Front track	57 in
Rear track	59 in
Curb weight	2980 lbs
Front/rear distribution	52/48
Test weight	3250 lbs
Turns lock to lock	3.6

Gear ratios:

Gear	*Standard*	*Optional*
3rd	3.55	3.27
2nd	4.65	4.28
1st	7.84	7.22
Rev	7.84	7.22

Tire size	6.70 x 15
Brake lining area	158.0 sq in
Fuel capacity	17 gal

PERFORMANCE

TEST CONDITIONS

40°F, light wind, dry concrete surface at sea level.

SPEEDS IN GEARS

Gear	*True mph*	*(Car) mph*
1st	64	(60)
2nd	108	(102)
3rd	118.5	(110)
Best run	120.0	

ACCELERATION

Range	*Time, Seconds*	*Gears Used*
0-30	3.4	1st
0-40	4.6	"
0-50	6.0	"
0-60	7.5	"
0-70	10.0	1st, 2nd
0-80	12.5	" "
0-90	15.8	" "
0-100	19.3	" "
50-70	4.5	2nd
50-70	5.8	3rd
60-80	4.7	2nd
60-80	7.0	3rd
Standing ¼ mile	15.9	1st, 2nd
Speed at end of quarter	91 mph	

FUEL CONSUMPTION

Hard driving .... 12 mpg (tank mileage)

View shows sleekness of new Corvette. Except for phony air-scoops and knock-off type discs, the car is functional in design.

Instruments are well balanced for eye-appeal, but not practical. Reading them at any speed over forty is difficult. Note simple design of steering wheel.

Small luggage compartment makes long trips restrictive. Jack fits inside spare to conserve space. Larger trunk would destroy line, increase weight.

Fully automatic, the top slips out of the well after the lid raises. Convenience of automation makes added weight of unit bearable.

Corvette interior is well appointed, with leg room for the six-footers. Wrap-around windshield makes getting in and out a bit awkward.

at the bottom of the dash, even if you can take your eyes from the road long enough to find them. The speedometer is well-placed, but quick correlation between the numerals and the divisions is impossible, as they are on different planes. Dial lighting is very good, with rheostat control, and the interior lights are perfect for rallying, being placed under the cowl.

Other interior shortcomings are the dearth of storage space, save for the between-seat compartment, and a conflict between the heater and the passenger's feet. The view forward is very impressive, and clever psychologically. The hood bulges, long fender lines, and cowl vents (which, incidentally, can easily be made functional for dry climates) combine to give an impression of great forcefulness. Vision over this snout is adequate, but not outstanding. The heater and defroster are well up to their jobs, and the only other irritant might be a very awkward and stiff interior door control.

In spite of numerous open car details, the passengers can be kept warm and dry, and can set their own climate at a literal touch of a button. I can imagine no greater contrast than between this and the forced exposure of the gutty old J2 Cadillac Allard, but the fact is that such an Allard in stock trim would be left behind at the quarter by this incredible Corvette! The figures speak eloquently for themselves, and with the lower 3.55:1 ratio things should happen even more rapidly. As a matter of fact, our speedometer was so very slow that it probably was geared for use with that ratio. Also, the engine was nothing like wound out at the top end, and the lower gearing would probably improve top speed by five to seven miles per hour.

SCI Road Test: Corvette

Precision balancing after assembly may have accounted for the clean, smooth running of the engine, and its ability to rev freely to around 5500. For about the first half inch of throttle travel only the rear one of the two four-barrel carburetors is working to prevent overcarburetion at low speeds. When the front quad cuts in the previously unobtrusive exhaust note sharpens and the car starts to move. When backing off at higher speeds there is a not unappealing rap from the duals. Idling is at 1000 rpm when the automatic choke is working and 600 rpm when warm, and the powerplant is tractable enough to lug down to a 12 mph in high.

Featuring special cooling and nine coils instead of the old diaphragm spring, the clutch took a lot of punishment without complaint. It is not easy to get a potent car off the mark with such very high gearing, and this component took the brunt of the effort without signs of heating or slippage. The gearbox wins similar praise for its well-chosen ratios and effective synchromesh. Shifting linkage is smooth and direct, the heavy-knobbed lever being spring-loaded to the right-hand side of the conventional "H" pattern. The synchro can be beaten by a very quick move from first to second, but the movement between the two top gears is impeccable. Synchromesh on low would be a useful boon, but a noiseless downshift can be made by double-clutching.

Due to the high ratios, the standard-shift Corvette is not really at home in town, and Powerglide might be better for urban use. Out on the road, though, as second gear takes over from first at around sixty and keeps the seat in your back 'til over a hundred, you learn what this car was made for. Cruising is effortless at 85 or 90, though with enough wind noise over the soft top to render the radio unintelligible.

It is in the handling department in particular that the Corvette proves itself the only true American production sports car. The steering is far from perfect but it is fast enough to allow right angles to be taken without removing the hands from the wheel, and this virtue will make up for many vices. The latter include an inch and a half of play, beyond which a strong caster action gives the wheel a springy feel. This little "no-man's-land" in the middle causes some trepidation in tight spots. Once the wheel has been set for a bend, and the car has assumed an initial roll angle, the steering and throttle response are fast and consistent enough to allow very precise control.

Like most American cars this Chevrolet is a very strong understeerer, and requires a lot of helm to keep it on line in a bend. The stock rear end damping is a little weak; too much so to make a full-blooded drift a stable proposition. Cornering speeds and behavior were markedly improved by tire pressure five psi higher than the standard of 25 psi front and 27 psi rear. Raised pressures plus stiffer rear shocks could combine with an already broad track, good weight distribution, and low center of gravity to make the Corvette a real fiend on corners. These criticisms, it will be noted, are minor, and apply equally to many imported machines.

Of course, tire squeal is not entirely absent during these high-speed direction changes, but the car stays in the corner so there can be no real complaints. The Corvette is at its best on a winding open road, and, like the Jaguar, is dramatic but uncomfortable on a twisty back lane. The test car would have been much handier there if the driver had had more arm room and the optional seat belts. He tends to be thrown around more than necessary, but is not as conscious of the car's roll angle as is the passenger.

Brakes are still by far the weakest link, and it must be admitted that they faded almost into oblivion during the performance tests. They recovered very quickly, though, and pulled the heavy car up with a minimum of slewing even when very hot. I sincerely feel that the substitution of harder Moraine linings or some of the foreign competition brands will improve high-temperature durability and perhaps modify the present spongy feel of the pedal. No power booster was fitted, and required pedal pressure was on the high side.

Very sensibly for a high speed car, the hood is hinged from the front, and opens well out of the way. The battery and brake master cylinder are easy to reach, and the small air cleaners ease access to the engine as a whole. Most awkward feature is the shielding for the ignition wiring, necessary to eliminate radio interference in a Fiber-glas body. Wingnuts quickly free these shields and bare the double-breaker distributor and all but the two left front spark plugs, which are tucked in behind the steering box.

The hydraulic system for the top mechanism is powered by a separate electric motor, which allows operation with the engine off and avoids direct absorption of any engine power. Individual motors operate the door windows.

Well finished and fitted, the trunk is usefully large for a sports car. All luggage must be removed to extricate the spare from its wooden-lidded compartment, which also houses the jack. One carrying feature of many imports that is missed in the Corvette is that handy space right behind the seats for coats, hats, lunches and other items that you don't want to store in the trunk. In the Corvette you either live with them or lock them away.

In almost every respect, the 1956 Corvette is a very satisfying car on the highway, and supplements astonishing performance with a high level of roadholding. Even as it stands, power equipment and all, it has become a serious competitor for Jaguar in Production Class C, and this is by no means General Motors' highest goal. In international events this year the car will be equipped with an optional cam providing 250 bhp at the sacrifice of present low-end smoothness. Also on the fire for either this or next year are engine boosts to 275 bhp, extensive use of light alloy in both body and chassis, and the development of suitable disc brakes by GM's Moraine Division.

It seems likely that the standard Corvettes will remain much as they are, with work on the competition versions proceeding simultaneously as has been the case with Jaguar and their C and D models. Another two or three years could probably see a racing Corvette with as many standard parts as the D retains from the Jaguar line, shrouded in advanced coupe bodywork. GM will learn an incalculable amount from these cars, much of which will be passed on to the standard Corvette and to the passenger car. They've already learned quite a lot, as a matter of fact, most of which shows up in the all-around excellence of the 1956 Corvette. #

How fast is the Corvette?

By ROGER HUNTINGTON

WHAT is this *stock* Corvette? After all, 150 mph and a standing mile in 39.6 seconds *on sand* . . . that's Ferrari or Jaguar performance, not the kind of go you expect from penny-pinching American production-line stuff. One can picture everything from fuel injection to a stroker kit under the hood. At first blush, the term "stock" seems just a shade far-fetched. But only at first.

I've done a little nosing around and dug up a few facts (and probably a little fancy!) on the question of Corvette performance. I'm still not prepared to *guarantee* anything in regard to the stock status of the 150-mph factory Corvettes, but here's the story for what it's worth . . .

In the first place, nobody who knows hot road machinery today will argue the fact that the new Chevrolet V-8 engine produces more horsepower and torque *per cubic inch* than any American production engine in history. At 265 cubic inches, it will actually out-perform many OHV V-8's with well over 300 inches. We don't need to take the word of a factory dynamometer operator for this. An afternoon at any drag strip, or a few runs against a stop-watch, will show quickly that here is one of the hottest production engines in the world — *regardless of piston displacement.* I have even cross-checked by road-testing with an accelerometer. Here we use the mass of the car itself as our dynamometer, read the actual rate of acceleration at various speeds on this instrument, and with these figures the horsepower curve of the engine as installed in the car can be calculated with fair áccuracy. Here are some eye-opening comparisons, both calculated from personal accelerometer tests on strictly stock, un-reworked cars with no special tuning:

'56 Packard V-8 in Studebaker Golden Hawk (rated 275 hp):

Maximum hp: 215 @ 4400 rpm
Maximum torque: 320 lb.-ft. @ 2500 rpm

'56 Chevrolet Corvette (rated 225 hp):

Maximum hp: 220 @ 5400 rpm
Maximum torque: 225 lb.-ft. @ 3400 rpm

(Admittedly, this was an exceptionally hot Corvette engine — but it was definitely stock. The average peak hp would probably be nearer 200.)

LEFT: Corvette team at Daytona. NASCAR allowed use of the Duntov competition cam. BELOW: At Sebring, Corvette's high torque placed it first at start, but couldn't keep apace.

Firestone "Bonneville" tire used by Corvette at Daytona. Note thin tread layer and smooth side casing.

Now I don't know why the Chevrolet engine performs this way . . . and I don't know anyone who does. I've talked to engineers, hop-up experts, and competition men; no one has any half-way reasonable explanation for the unusual performance. It's as much a surprise to the Chevrolet engineers who *designed* the thing as it is to the people who are out beating the big-inch devotees with it. The engine was designed primarily to be mass-produced for a few dollars. By some crazy trick of fate it *goes.*

So let it be firmly established at this point that we do definitely have brute horsepower available in the '56 Corvette — not just advertised horsepower.

And there's another angle to this horsepower problem: Several weeks before Daytona Chevrolet's engineer, Zora Arkus-Duntov (of Le Mans, Allard, and Ardun head fame), designed a special "competition" camshaft for the Corvette engine. It has approximately the same timing and lift as the standard cam, but much faster opening and closing rates. Official dynomometer figures on the combination have not been released; but Duntov says the peak of the curve on it will average about 240 hp at 5800 rpm. Maximum torque is slightly below the standard 270 lb. ft., but comes in at 4400 rpm instead of 3600. Approximate power curves for the two cams are shown in the accompanying graph. The new cam is characterized by a very flat peak in the 5000-6000 rpm range, without a radical drop-off in power at low speed.

This competition cam was in all the factory Chevrolets at Daytona — and the deal apparently satisfied NASCAR stock class rules requiring 100 cars to be consigned to dealers prior to February 1st. It is optional on all Corvette engines now, including those installed in sedans. Ask for the Duntov cam — (but don't try to balance a coin on the hood when it's idling!).

Then there was the well-known modified Corvette at Daytona. There were all kinds of wild rumors about what this had in it. Actually, Duntov tells me this move was merely to let Corvettes compete in one additional class in the Speed Week festivities, or the class for modified production sports cars. The only change was a set of experimental cylinder heads with 10.3:1 compression ratio. They added about 15 hp to the peak, or something around 255

Three of the four Corvettes entered in Sebring had standard 240 hp engines shown here.

German ZF four-speed gearbox was used in the modified Sebring machine. Extra gearing helped place car in first 10.

Of the four Corvettes entered in Sebring two finished: number 1 driven by Fitch and Hansgen, and number 6 above.

Corvette's brakes were over-sized cerametallic Bendix units. Primary shoe has two bonded surfaces, second shoe four.

Three-quarter rear view of the Corvette at Daytona. Faired headrest fin increased top speed by 1½ miles an hour.

Stock Corvette power plant produces more horsepower per cu. in than any American production engine in history.

"...here is one of the hottest production engines in the world - regardless of piston displacement."

BELOW: *Among other changes the Sebring race Corvette was equipped with Halibrand magnesium knock-off wheels.*

Last minute details are checked just before Sebring race. Number 1 car had 316 cubic inch engine for class B.

hp. I have no reason to doubt that this was the only modification, as there was very little difference in times between this modified and the standard Corvettes at Daytona.

Of the four factory Corvettes sent on to the Sebring 12-hour race, three had the standard 240-hp engine described above. Other changes included Halibrand magnesium knock-off wheels, special shocks, and special oversize brakes (wider, same diameter) with Bendix cera-metallic aircraft linings. (All this is to be optional equipment). The fourth Sebring car had all the above, plus a 4-speed gearbox and special 316 cu. in. engine to put it in Class B. Very little information is available on this car. Duntov laughingly refers to the engine modification as a "California tune-up," but wouldn't give any details. The implication, of course, is a big bore and stroke job, additional cam, complete cylinder head reworking (oversize valves, higher compression, porting, polishing), and special ignition. In view of all the evidence, a conservative output estimate would have to be near 300 hp at 6000 rpm.

The two other key factors in straight-line performance are weight and drag. Weight-wise, the '56 body modifications didn't help matters. The curb weight of last year's V-8 Corvette ran around 2800 lbs. with soft top. Current showroom models are up about 150 lbs., due mostly to the wind-up windows and power top-raising mechanism. The stripped competition Corvettes will curb down around 2700.

Drag is a more important factor than weight when we're after top speed. Open sports cars have a serious wind resistance problem because of the larger suction area behind the upright windshield. Such a car will almost invariably go three to 12 mph faster with the top up than with it down — and frequently a small sports car with the top down will prove to have as much total air drag as a medium-size sedan. (Accelerometer tests substantiate this.) The best solution, of course, is to whip the windshield off altogether, give the driver just a narrow, low screen for wind protection, and cover over the passenger side of the cockpit. This not only reduces the drag coefficient by eliminating the big suction area, but frontal area is also substantially reduced.

This is exactly what they did for the competition Corvettes. No wind tunnel tests were run to determine the improvement, but I would estimate that total air drag at any speed is reduced 40-45% with this layout, compared with a standard Corvette with top down. This streamlining kit is supposed to be available as optional equipment.

Tire rolling resistance was another vital factor in the top speeds made by the Daytona Corvettes. They would have been lucky to reach 135 mph with the factory-equipped tires. Few of us realize that it requires about 160 hp to roll these tires at 150 mph – under the weight of the Corvette and inflated to 28 lbs./sq. in. Rolling resistance is caused by the violent flexing and stretching of the tire casing and tread layer under rolling and centrifugal forces. We can reduce it by increasing inflation pressure, using stiffer 6-ply casings with a steep cord weave angle, and with a lighter, thinner tread layer. The special Firestone Bonneville tires have the lowest rolling resistance factor of anything available in this country, and they compare favorably with the European super-speed types for record cars (Dunlop, etc.). They have only a fraction of the drag of regular passenger tires, and seem to be substantially below the conventional full-tread racing tires like the Firestone Super Sports 170 and Goodyear models. Most of the really fast cars at Daytona, including the Corvettes, were on Bonneville rubber for their top-end runs, running inflation pressures anywhere from 40 to 65 lbs.

Add up all these drag-reducing gimmicks – stripped windshield and competition screen, cockpit cover, partially masked grille opening, headrest fin, Bonneville tires at 50 lbs. pressure – and it would take roughly 190 hp at the clutch to move the competition Corvette 150 mph in still air on pavement. So you see, things begin to fall into place. The absolute top speed potential of this combination on pavement, using the 3.27:1 rear-end ratio, is definitely *over 160 mph.* Duntov clocked 163 mph at 6300 rpm on the Arizona test track. An honest 245 hp should account for this – which appears to be well within the capabilities of the engine in stock form, with options.

The best two-way average on the beach was 150.53 mph. Duntov feels the true potential here is over 155, given excellent beach and wind conditions. Rolling resistance on the sand is definitely higher than on hard pavement – and, as beach conditions deteriorate, resistance goes up and traction comes down. Also, wind is a problem at Daytona. *Any* kind of wind component parallel to the course acts to chop down the two-way average; the wind holds you back more on the upwind run than it boosts you downwind. Coast-wise winds of 35 mph are not unusual at Daytona.

Conditions weren't perfect on any of the runs. On his 150 mph runs prior to Speed Week Duntov had little wind, but only a fair beach; he was getting five percent wheel-spin. On Fitch's official two-way runs with the production Corvette during Speed Week the beach was good, but there was a 20 mph coast-wise wind. He turned 155 mph downwind, but the two-way average came out to 145.54. Duntov averaged 147.3 mph in the modified class that day. He figures if he could have switched to a 3.08:1 axle ratio for the downwind run he could have increased his speed from 156 mph to more than 160. At any rate, it appears that the true top speed *potential* of the competition Corvette on Daytona Beach is a good five or 6 percent below the speed on hard, smooth pavement. That should answer some burning questions!

I have listed a set of road test figures which I obtained personally on a friend's strictly "showroom" '56 Corvette . . . then for comparison I've *calculated* the corresponding figures for the Sebring Corvette, using slide rule, graph paper, and a planimeter (area measuring instrument for graphical analysis). I think I hit the true performance of the Sebring cars pretty close with this method, by using all available technical data on the cars and carefully estimating the "unknowns."

Incidentally, the calculations on this car are based on a gross weight of 2900 lbs., 3.55:1 axle ratio, and a shift point of 6500 rpm in the gears. (This was used by Duntov on his standing mile runs, but might be a little high for a long race.) The showroom Corvette had stick shift, 3.55:1 axle, but not the Duntov cam. It seemed to run well in all rpm ranges, but there was no special tune-up for the test. It had about 700 miles on the odometer.

Here are the figures on the two cars for comparison:

	Sebring	*Huntington*	*SCI Test*
0-30 mph. . . .	3.2 secs.	3.2 secs.	3.4 secs.
0-60 . . .	6.3 secs.	7.5 secs.	7.5 secs.
0-80 . . .	9.6 secs.	12.4 secs.	12.5 secs.
0-100 . . .	15.1 secs.	21.2 secs.	19.3 secs.
Standing 1/4-mile . . .	14.9 secs.	16.0 secs.	15.9 secs.
Speed at 1/4 . . .	99 mph	89 mph	91.0 mph
Top speed . . .	Approx. 148 mph	(top speed) 121 mph	120.0 mph

And there you are. There's a pretty big difference between the performance of our production-line Corvette and the Sebring job, but obviously both of them can hold their own in pretty quick company. The showroom job would have strictly no trouble with its arch-rival, the showroom T'bird – and the Sebring cars, I think, might even raise a few eyebrows on the California drag strips.

No doubt about it . . . the new Chevrolet Corvette is America's fastest stock car. #

The formula is shorthand for the most significant advance yet recorded in American sports cars. It means: The 1957 Corvette V8 with fuel injection turns out *one horsepower per cubic inch of displacement*—and there are 283 cubic inches on tap!

To anyone who knows cars, that fact alone is a warranty of magnificent engineering. But the driver who has whipped the Corvette through a series of S-turns really knows the new facts of life: This sleek powerhouse *handles!* Matter of fact, you can forget the price tags and the proud names—no production sports car in Corvette's class can find a shorter way around the bends!

Post another unique item: The Corvette is a *comfortable* sports car! Wind-up windows, deep-cushioned bucket seats, optional hardtop—even Powerglide* if most of your driving is in tough traffic. Power-Lift* windows, too. (Competition drivers use them because they weigh *less* than the manual mechanism.)

Seriously, if you haven't punched the throttle on that fuel injection jewel, you're missing the king-size thrill of the American road. That ought to be corrected—and soon. Let your Chevrolet dealer check you out. He's always glad of an excuse to drive a Corvette! . . . Chevrolet Division of General Motors, Detroit 2, Michigan.

**Optional at extra cost. Powerglide available exclusively with 220 h.p. standard engine or 250 h.p. fuel injection V8.*

A big one-step boost for the Corvette is given by the Latham axial-flow blower, belt-driven and fed by two Carter carbs.

By ROGER HUNTINGTON

HEATING UP THE HOT ONE

"THERE'S no substitute for cubic inches."

That's a popular theory among American auto enthusiasts which eight years of bigger and bigger passenger car engines have done nothing to kill. That is, until the little 530-pound, 265 cubic inch Chevrolet engine came along. After two years of experimenting with this fireball, a lot of smart hop-up men aren't so sure about the pet cubic-inch theory. They're beginning to realize what the Europeans have known for years—that a good, big handful of *usable* rpm can replace a lot of inches.

This is the double whammy that the Chev engine has on the other V-8 powerplants throughout the industry. It has about another 1000 rpm of usable crank speed that can be turned into brute horsepower at the road wheels. Furthermore, it's considerably smaller and lighter than the competition. It's an ideal powerplant, either for hopping up in a stock Corvette or Chev passenger car or for transplanting into other sports cars and specials. It will be easier to get one horse per cubic inch out of this mill than anything else you can lay your hands on this side of the British and Italian sports car factories.

Here's how . . .

FIRST THINGS FIRST

Before we go any further, take note that the Chevrolet V-8 must not be considered a "big, high-torque Detroit engine." So many sports car enthusiasts, used to thinking in terms of one to three liters, look upon all American V-8's as big, hairy iron with enough low-end torque to pull the skin off an elephant. The Chev is not in this category. Cubic inches are the prime factor in low-end flexibility and peak torque—and the Chev just doesn't have 'em. Even when bored and stroked near the practical limit—around 320 cubic inches, it isn't big by Detroit standards. So it might be good to try to picture the Chev V-8 somewhere in between a 350 cubic inch Detroiter and a two-liter European sports engine. In other words, we can profitably increase displacement with an eye to improving torque and flexibility, but we're never going to get a "top gear" powerplant. Modifications aimed at utilizing the high rpm potential to develop *horsepower* will pay off more.

By the same token, this engine is always going to want use of the gear lever. A good "bread and butter" rev range for quick road travel is 3500 to 5500 rpm. With proper modifications the 4500 to 6500 range can be used for com-

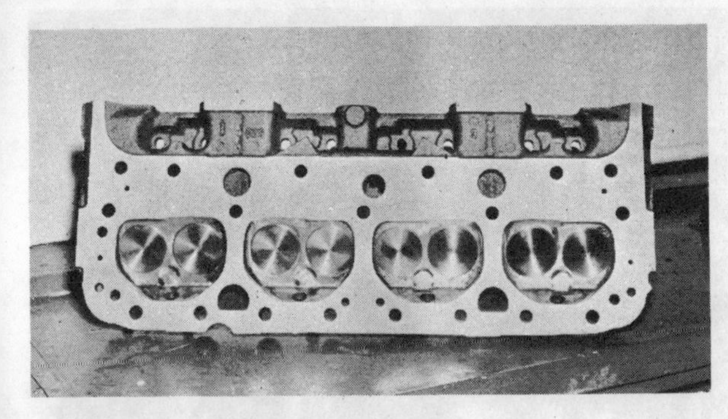

Polishing Chev chambers and lightening valves is worthwhile. These intakes are stock, while exhausts were recut to take 1.625 inch valves.

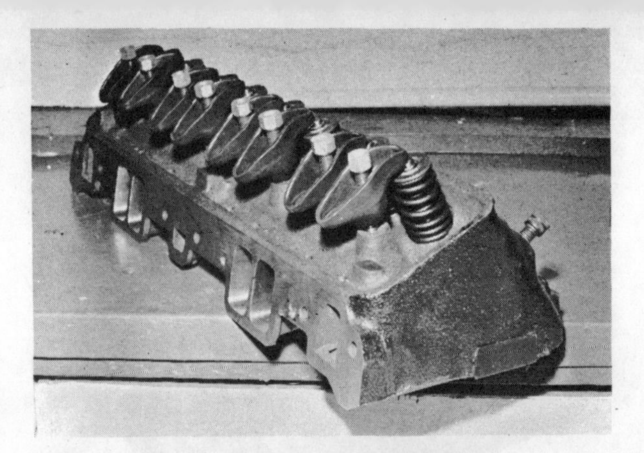

Credit for rev-readiness is often given to light, simple stamped rockers. Double valve springs, rated at 200 pounds, are final touch. Ports are smooth.

Exhaust ports can be hogged way out. These were once square! Valve guide bosses are kept healthy, for stem protection. Allen cap screws hold manifold down.

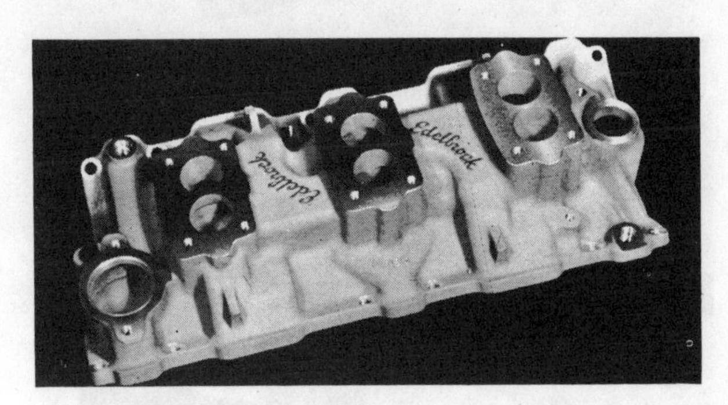

Best carb setup for Chev so far is Edelbrock's 180 degree triple-twin-throat manifold. New progressive throttle linkage keeps it tractable on the street.

McGurk's special Chev valve springs give 260 pounds pressure with valve fully open.

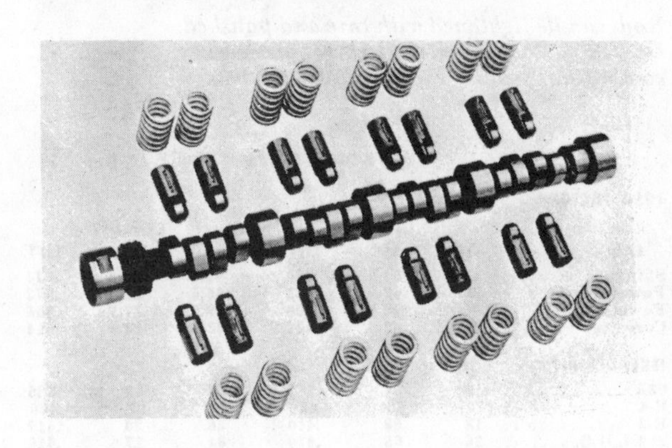

All-out versions usually justify added cost of roller cam kit. Weber set is distinguished by slot-type tappet location. Broad lobes don't always mean wild timing.

petition—and this can be extended under some conditions. All this means that items like head reworking, plentiful carburetion, hot cams, light valve gear with proper spring tension, etc. will probably give more results *per dollar* than boring and stroking.

Let's look over these categories in more detail:

BORING AND STROKING

The '55 to '56 block, with 3¾ inch bore, can be safely overbored a maximum of ⅛ inch. The '57 block, with 3⅞ inch bore has been recored to give, in effect, thicker cylinder walls (with the same bore center-to-center distance). However, the new walls are not quite so thick, so about .100 of an inch would probably be a safer maximum overbore. A block with very little foundry core shift can be bored to four inches.

Pistons for these large oversizes are widely available from a number of California supply houses—Speedomotive, Turner, J. E., McGurk, Venolia—and can be had for almost any desired stroke length. These can be ordered as heavy-duty racing pistons, with wide, noisy clearances and some additional oil pumping, or as a milder, less-expensive road type. You should consider "pop-up" domes as a simple, fool-proof way of increasing compression ratio without the problem of head milling. Most of the piston manufacturers will put domes on your pistons for any desired compression, and they will also fly-cut for proper valve clearance. Be sure to specify bore and stroke, type of cam, etc. Prices for a set of these special pistons, including rings and fitted pins, generally run from $60 to $100.

Stroking, as you know, is the practice of building up the crankpins by arc welding and then grinding them back down to size on a center that is offset away from the crank axis a distance of one-half the desired stroke increase. About the only limit to the stroke length obtainable this way is interference between con rods and cylinder walls (though some engines will foul cam lobes). Chev V-8's have been stroked ½ inch, though ¼ inch is more usual. A stroke increase of .22 inch in conjunction with a 3⅞ inch bore will give 304 cubic inches up to the limit of Class C modified. The '56 Sebring modified Corvette had a .25 inch stroke to put it in Class B (307 cubic inches). The crankshaft companies like Ansen, C. & T., etc. usually sell a stroked crank as part of a complete kit, including pistons, rings, pins, con rods, and bearings. The whole assembly is carefully balanced before

If you don't trust those stampings and studs, this shaft kit with cast magnesium rockers might be the answer. Thomas makes it, to bolt into old stud holes.

One of goingest Chevs around, the Carstens HWM engine was dyno-tuned and fully equipped at Edelbrock's shop.

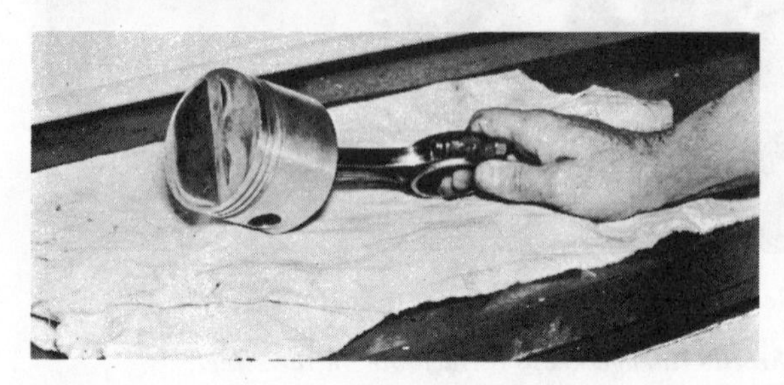

Rods can be lightened with care and polished. Domed section on full-skirt JE piston raises compression, while "eyebrows" clear valves.

Factory "Ramjet" fuel injection is more for publicity than power. Nozzles are long way from ports, and act as small carbs at idling. Fuel metering is accurate.

CHEVROLET CAMS

CAM	INTAKE BTC	INTAKE ABC	INTAKE LIFT	EXHAUST BBC	EXHAUST ATC	EXHAUST LIFT
1956 FACTORY						
Standard	18.0	54.0	.334	52.0	20.0	.334
Powerglide	26.5	63.5	.373	66.5	23.5	.373
Powerpack	31.7	92.5	.396	69.6	37.0	.396
Corvette	21.5	62.5	.404	62.5	23.5	.414
ISKENDERIAN						
02A	12	54	.385	52	14	.385
E-4	20	60	.400	58	22	.400
E-2	18	58	.410	56	20	.410
#2 L.D.B.	26	65	.410	64	27	.410

shipment. This seems to be the best, most reliable way to go bigger—that is, bore a moderate amount and get the full stroker kit. Prices are steep—$300 or over for the works—but torque increases are very gratifying. Also, the balance job is a must for high rpm.

HEAD REWORKING

You can go shallow or deep into this job. The Chev V-8 already has generous valve sizes in relation to its displacement, so doesn't respond to diameter increases as well as some engines. Intakes are 1.72 inches head diameter, and exhausts run 1.50 inch. One simple trick that will help, though, is to enlarge the port diameter *under* the valve. Production engines have valve seats over 1/8 inch wide (a lot wider than is needed to seal the cylinder) to allow for machining errors and core shift. It's easy to sink a 70 degree reamer down into the port, using the valve stem bore as a pilot, to open up the port diameter until the seat is reduced to a width of around .050 inch (around the outside edge). You can bring the seat out still closer to the valve edge by grinding the seat itself with a 45 degree stone (also piloted in the stem bore), and then opening up the port still more with the 70 degree reamer. These cuts must then be blended into the head surface and port passage with a hand grinder. If desired, you can also turn some stock off the under side of the valve heads to take full advantage of the narrower seats (see drawing).

It is possible to get 1/8 inch larger intake and exhaust valves into this combustion chamber. The modified Corvette with which Arkus-Duntov cracked 150 mph at Daytona a year ago had 1/8 inch oversize intakes and exhausts made by Thompson Products. These special valves are not available for general distribution, so most of the boys use 1.88 inch intakes from '53 and later Chevrolet Powerglide six-cylinder engines. This has the same stem diameter (.3419 inch with .0019 clearance) and can be adapted by shortening to the length of the V-8 valve. Regroove it to accept stock spring keepers, and reharden the valve stem tip after cutting off. These valves have a 30 degree seat angle, so the outside diameter of the seat must be enlarged to about 1.85 inch with a 30 degree stone and then the port bored out with the 70 degree reamer to bring the seat width close to .050 inch.

Complete McGurk stroker kit gives 306 cubic inches, retails for $300. Fully balanced, of course.

306-inch engine on McGurk's dyno had works, showed 298 bhp at 5500. In sedan it turned 102 at drags!

High dome and deep valve cutaways are cast into special McGurk piston for big-bore Corvette. Resulting ratio of full-skirt slug is 10/1.

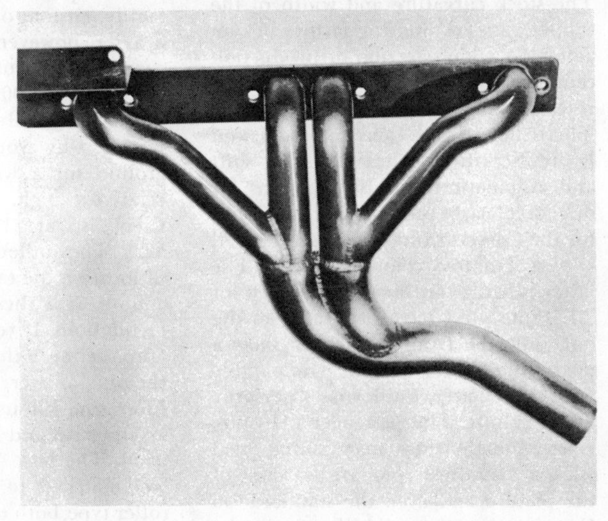

Hedman "Hedders" for Chev are clean and well-designed. They're particularly good with pre-1956 cylinder heads.

Exhaust valve diameter can be increased to 1⅝ inch by using late Lincoln V-8 valves; these have a compatible stem diameter (.3414 inch with .0024 clearance) and can be adapted by shortening and regrooving as above.

And don't forget to clean up the rough port surfaces with a hand grinder. Port walls are pretty thin on the Chev engine, so don't go too deep. The stock section areas are good anyway, and the goal should be a constant area. The later '56 Corvette engines came through with cast-larger ports, achieved by modifying foundry coring. If you have to buy a set of heads, look for these. Even if you don't want to go to the effort of doing a thorough porting job, it's always a good idea at least to *match* the ports in heads and manifolds.

Then there's the matter of head milling. The Chev heads don't have the meat in them that some have. A cut of .090 inch should be considered the absolute maximum. This will raise compression about 1½ ratios. A mill of more than about .060 inch on '55 heads and .020 inch on '56 Corvettes will cut into the valve seats and cause piston-valve interference, so the seats must be recessed for the very deep mill job. Again, pop-up pistons are the answer. A cut of .060 inch on the '55's will raise compression ratio one point, which is usually all you can handle. When doing any head milling on the Chev engine always check for piston-valve interference since it occurs readily and will vary with cam rates and lift. Most of the Corvette V-8 pistons, including the factory SR slugs for use with the Duntov cam, are fly-cut notched on the heads for valve clearance. They usually have small shoulders cast on the underside of the piston heads to allow somewhat deeper cuts if needed. Possibly the neatest way to increase the compression of a '56 Corvette engine is to bore ⅛ inch and use stock pistons. This would give 283 cubic inches and a compression ratio of 10 to 1. Similarly, the 9.1/1 '56 heads give a 9.5 ratio when clamped on the '57 block.

I don't recommend ratios much above this for the Chev V-8 on 96-98 octane premium pump gas. Its octane requirement in relation to compression ratio is not the lowest in the industry. A lot of fellows who have tried to use 11/1 for the street, as many are doing with other V-8's, were sorry. The '57 fuel injection jobs are able to use more compression because of the more even fuel distribution between cylinders.

Corvette

CAMS AND VALVE TRAIN

The unforseen rpm potential and top-end breathing of the Chev V-8 *in stock form* are already legend. Thousands of enthusiasts are asking if there is anything to be gained by putting money and effort into modifying this apparently fine cam and valve train setup. After two years of observation and after careful study of a number of dynamometer test results that cannot yet be made public, my answer to this question must be an unqualified "yes".

The manufacturers are grinding their own "full-race" cams these days but they still have to compromise. Hydraulic lifters cause a certain performance sacrifice as does a half-way decent idle. The stock curvature and width of the tappet base are limiting factors in cam action. A wide selection of production cams has become available for the Chevrolet V-8 since its introduction in 1955. There have been cams for solid and hydraulic lifters, cams for stick shift and automatic transmissions, cams for dual quad carbs on passenger cars, cams for the Corvette and, of course, the well known "Duntov" competition cam. The latter is being continued unchanged for '57. From all I have been able to dig out, only the Duntov cam can show a peak horsepower output comparable with the better California "performance" grinds. The passenger car cams, even though they may sound real wicked over 5000 rpm, are way out of the picture and even the standard '56 Corvette cam (for solid lifters) may be as much as 10 or 12 percent off the better California kits using flat solid lifters.

Of course, there are many factors involved in cam selection. For one thing, there must be metallurgic compatibility between cam lobes and lifter faces to prevent rapid wear at high rpm. You need have no fears about this if you use factory parts, but some of the specialty cams can't be trusted under hard operating conditions. Another factor is valve float speed. If you use your Corvette much for competition it's very pleasant to have 1000 rpm or so in hand over the power curve peak, so you can play with the gears and downshift without being afraid of wrecking the valve gear. It has been found that float speed is heavily influenced by valve spring pressures and cam acceleration (on both flanks). The standard '56 Corvette cam can be counted on for only about 6100 rpm, when using the stock 160 pound spring pressure at the valve-open position. The California cams I have test figures for, even when using well over 200 pounds pressure, would vary anywhere from 5600 to 7800 rpm in float speed!

And there's a little matter of money. You can buy a Corvette cam these days over the counter for $18; it's a simple matter to collapse your stock passenger car hydraulic lifters—and you're in business with a "racing" cam. A good California flat cam kit, including compatible lifters, springs and possibly pushrods, can run you over $200. That's a lot to pay for a 10 or 12 percent increase in horsepower. At the same time, the Duntov kit lists over the counter at around $180, which isn't hay.

Finally, there's the question of the roller cam. Since a small roller is used for the cam follower here, cam contours are not limited in any way by lifter base width or radius; the sky's the limit on lift, rates, dwell, etc. A good roller cam should give substantially more peak output than any of the flat-lifter family, custom or factory. The lifter is heavier, however, so more spring pressure must be used to keep the float speed above 7000 rpm. Medium-speed torque is usually poor, but there's no reason why you couldn't have one ground for a wider range. Prices are relatively high. The Herbert kit for Chev lists at $175 and the Weber at $225. They differ mainly in the method of locating the tappet. So I'll go out on a limb with these specific cam recommendations. If you want to stay strictly "production", the Duntov is definitely the only answer. For a flat or radiused-lifter cam kit in the custom category, Iskenderian and McGurk products look good. The Isky "E-2" grind is specially well-matched to the Chev V-8. For a roller type, both the Herbert and Weber are good cams. The roller layout is preferable for any "all out" engine.

Modification of the Chev valve train itself is a very controversial subject. Some experts say the light weight and great stiffness of the unique stamped rocker arms and tubular pushrods are keys to the fantastic performance of the engine. These would greatly reduce deflection of the valve train, so the valve motion would follow the cam more closely, which, in turn, is a big factor in breathing. Could be. Others feel this valve setup is too flimsy for really high rpm work. Some have gone to great lengths to install complete rocker assemblies from other engines on the Chev head. Thomas has a kit selling for $95 for converting to a regular rocker shaft and cast magnesium rockers by threading the original Chev rocker stud holes to bolt in the new rocker stands. Very neat.

CARBURETION

The Corvettes have a lousy carburetion system for a sports car. A four-barrel carb has a wide fuel bowl, which makes it very susceptible to fuel starvation under centrifugal force on corners. *Dual* four-barrel carburetion is no cure-all, either. We have the same starvation problem, plus poor throttle response and mixture distribution at the low end. The manifold layout is not the best from the standpoint of breathing restriction at the top end and the usual linkage synchronization problems exist. The Corvette team had a foolproof non-starving dual-quad setup at Sebring '56 but you may not want to carry five electric fuel pumps around. If you must use a four-barrel carb, one is a better all-around compromise than two. It should show more torque below 4000 rpm, better throttle response, no linkage problem, and it should come within three or four percent of the peak horsepower obtained with two!

I prefer *three two-throat* carbs for the Chev V-8. Several California companies now have the necessary 180 degree manifolds, and any of them should show more peak bhp (other factors equal) than the stock Corvette dual quad setup. The Carstens HWM-Chev is a case in point. Throttle response is better, low-speed torque is better, the linkage problem not appreciably worse (though still a problem), while the smaller fuel bowls practically eliminate fuel starvation on turns. A new twist on this setup is Edelbrock's novel "progressive" throttle linkage. The center carb only is used for cruising; at some pre-adjusted throttle opening on this carb—maybe two-thirds open —the throttles on the two end carbs start to open at a faster rate, and all reach full open at the same point. Olds is now plugging a production version, rigged to open fully at the ¾ point.

Same deal with fuel injection. The stock '57 system is very clever. It gives excellent throttle response on the road, fair low-end torque, reasonably convenient street operation in all kinds of weather, a negligible horsepower margin over carbs, and no fuel starvation. It would probably be superior to any carb setup for production racing if only for its responsiveness. But for competition in the *modified* category it would be a poor substitute for the famous Hilborn racing injection system. This meters fuel according to throttle position only, so the air flow restriction of venturis and manifolds is eliminated. It's not yet suitable for the street, but it should show a peak horsepower margin of seven percent or so over the standard Chev f.i. system in competition trim.

Supercharging is another possibility for the Corvette engine. At present, both McCulloch and Latham have kits. When we get up in this 250-300 bhp range, the percentage boost given by these units is not the 40 percent claimed, but more like 20 percent. This is because the *air flow* capacity and not the pressure capacity of a blower is limited; the engines keep getting hairier each year while the blowers mark time. For big stuff like the Cadillac and Chrysler engines, the little blowers are gasping for air. They will, however, still provide a useful boost for the Chev engine.

EXHAUST

The stock Chev exhaust manifolds can't compare to streamlined headers welded out of steel tubing—which actually will give a mild *extractor* effect. Several California companies supply these headers for Chev passenger cars, but, as far as I know, demand for Corvette units has not yet been sufficient to warrant tooling up. Custom installations can be had for less than $100, however. The popular Hedman kit fits the '55 head ports well, but is a trifle small for the enlarged 1956 version.

IGNITION

Stock Corvette ignition will do an adequate job in the usual speed range up to 5500-6000 rpm. When breathing is substantially improved by the foregoing modifications, things get more touchy. A hot 30,000-volt coil (Bosch, Mallory, D.S.M., etc.) is the first step. Then replace the entire ignition system with one of the high-quality custom sets like the Spalding "Flamethrower", Mallory "Magspark", or Jackson "Roto-Faze". These are all excellent ignitions, and should handle any load or rpm you throw to them. Magnetos are no longer considered essential on all-out engines, due to the tremendous improvement in coil ignitions in the last four years.

All the above changes will throw heavy loadings on the bottom end of the Corvette engine, which so far seems to be healthy in this department. The lively response and revvability of the Chev may also be due to the fact that its crankshaft, at 47.75 pounds, is the lightest in the industry by three pounds. The 19.02 ounce rods are also the lightest than can be had. This all spells low rotational inertia, or less shafting weight to be accelerated—like cutting down the flywheel.

With these points and provisos in mind, you can jump into the racing fracas with a power package that isn't awed by anything with a foreign accent. In a light, maneuverable chassis the Chev will nibble hungrily at any and all opposition, irrespective of price FOB Modena. With experimenters now working on single-overhead-cam conversions and Bosch injection installation, the limits are still way out of sight. This is the "flathead" of the future, men!

—Roger Huntington

Fuel injected Corvette takes in a tight, test bend. Cornering and roadability were of usual Corvette caliber. On this curve, car is a wee bit bent; sticking but building heavy loading on outside rear tire.

SCI DUAL ROAD TEST

DUAL QUAD AND INJECTED CORVETTES

Aside from fuel injection, this Corvette looks much the same as others. There were two differences in the two cars—The injected Chev had hardtop, and the carbureted model had a soft top.

CHEVROLET'S injection is a premature baby, but it's still alive and kicking. It was prematured by a sudden jolt from the collective Plymouth and Ford styling departments, and without a major body change Chev needed a potent sales weapon. The decision to bring out fuel injection was made very, very late in 1956—virtually on the introduction deadline.

At that time it was still what engineers call a "breadboard layout"—a combination of units that work properly, but aren't fully developed and integrated into one mechanism. It's still very much that way. Moreover, the Rochester engineers are very competent carburetor designers, but didn't have nearly enough time to adapt their techniques to the wholly different problems of injection. In spite of the fact that the new arrival had long been expected, then, its arrival was not without complications.

The above are facts, but in spite of them the Chevy "Ram-jet" constant-flow injector works very well indeed. The big question, of course, is: How does it stack up against carbs? To find out, we set up a full transcontinental '57 Corvette road test, involving our entire testing staff. On the West Coast an injected machine was obtained from the factory representatives, while on the East Coast a dual-quad version was provided by Alvin Schwartz Chevrolet and Shelly Spindel, of Brooklyn.

As a result we have a lot of the answers, but not all, the doubt being due to slight variations between the cars tested. the injected car, with the hardtop alone, weighed 120 pounds less than the carbed machine, which was toting the more complex soft top mechanism. Also, rear end ratios were 3.70 with the injector and 3.55 with the carbs. Advantage is taken of the better fuel distribution of FI by upping the compression ratio to 10.5/1, while the standard car had 9.5/1. This is a logical follow-up step, and does not "favor" the injection so far as our tests are concerned. The same can't be said of the weight and rear end discrepancies, which would tend to help the FI car at the bottom end, though very slightly.

Starting the dual-quad car was easy, by twisting the ignition switch, though some care was needed to avoid flooding on hot starts. Once warmed up, the idle was low enough at 500 rpm, but it was full of lumps and shook the car bodily. This can be handed to the competition cam, which was installed in both cars and checks out as follows:

Intake opens	35° BTC
Intake closes	72° ABC
Exhaust opens	76° BBC
Exhaust closes	31° ATC
Lift	.398 inch

The power from this cam comes on strongly at about 2700 rpm and stays that way until about 5300, after which it falls off rapidly, apparently due to valve gear. At the end of a

Corvette will drift in 75-80 mph range, as here, using power for control high up in second gear. In this case however damp concrete surface left little or no margin for error. Behavior could be predicted if rear end were stiffer.

Spare lays out of harm's way under trunk mat. Luggage space is adequate for small pieces.

Dual quad carb setup. Acceleration difference between this and the injected job were slight.

Cockpit layout. Cushioning is firm. Small leg room for medium to tall men. Seats are sat on, not in. Photo by Wilson

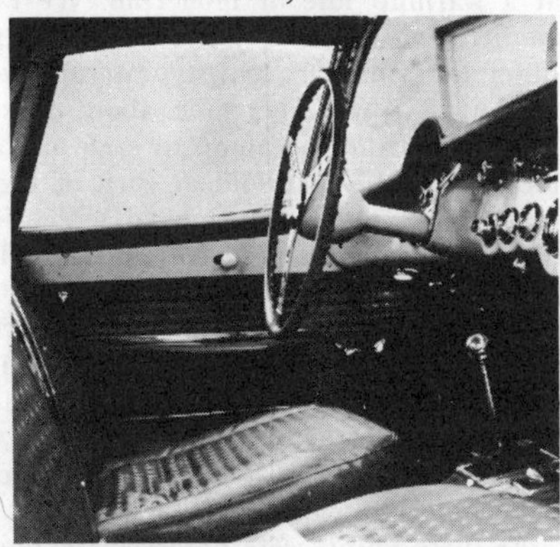

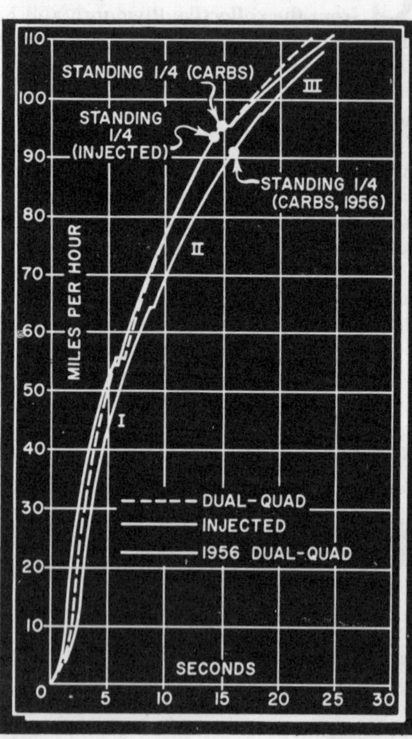

Low, wide and grinning grille is still mark of Corvette, injected or carbed. The soft-topped, dual quad car above weighed 120 pounds more than the West coast job, and carried a 3.55 rear as opposed to 3.70.

fast run, the idle was extremely bad, and after each stop in the braking test the carbs would stall the engine dead. The dual-quad setup is by now a familiar one, so rigged that the rear carb runs all the time and the front one cuts in only at about ⅔ throttle.

The engine was cold when we first twisted the switch on the Ramjet Corvette, but the engine caught on the first spin, bursting into a wholesome clatter of solid tappets jingling at a warmup idle of 1400 rpm. We raised the hood and studied the injection system. It looked purposeful but very different from the racing-type setups we know best . . . at present. A tiny copper tube, about the diameter of a pencil lead, runs from the pump to each of the nozzles which are located on the manifold at each of the intake ports. The tubes look fragile. We looked with distrust on solenoids, diaphragms, flex-cable drive, and complex linkages each of which might be a point of failure.

The single throttle body and venturi are not in accord with American ideas on racing injection, but resemble both carburetor practice and overseas applications. It's no handicap, as long as the orifice is big enough, and as long as ram intake tubes can be accommodated, and it gives very good metering for a wide operating range.

As these points were observed the engine coolant rose to about 170 degrees F. and the idle dropped to a mere 1200. We had occasion to switch off, and the next lesson in FI came when we tried to restart. Nothing happened; the starter ground on and on and the engine didn't even cough. So we tried treating it like a flooded carbureted engine and put the throttle on the floor. The engine caught within three spins. The remainder of our test bore out the fact that, while Chev's FI has the cold-start problem whipped, hot starts are rugged unless you use this simple trick.

Our test car's 1200 rpm idle seemed excessive and we experimented with this adjustment which is utterly simple and consists of a spring-loaded screw in the throttle body. Playing with this adjustment is not remotely critical and it is easy to obtain a decent idle of 450 to 500 rpm which is identical to the carbureted car. This casts interesting light on the controversy of continuous-flow injection versus timed-flow, Chev's being the former. However, there is so little torque at this rev level with the competition cam that it is very easy to kill the engine. Experiment showed that about 950 rpm is an all-around good idle speed. With this rapid an idle, you must first briefly engage high gear in order to slip silently into low when standing still.

On both cars, the action of the throttle linkage itself left a lot to be desired. The accelerator pedal travel is so short that you can't feather these potent engines, and the action is erratic and plagued with lost motion. In the bargain, the dual quads are at their worst at the bottom end. As a result the throttle response of the standard Corvette was unpredictable and slow, though things do happen in a big way after it takes a deep breath. This was still a handicap to clean control and rapid downshifting.

In sharp contrast, and making allowance for the tiny linkage, the throttle response with FI was instantaneous. It's as though there's a system of levers between the throttle pedal and your back. Press the throttle a little bit and your back is pressed, NOW, to the same extent. Slam the throttle down and your back is slammed, NOW, even in high gear. Second cog is so close to high that the slam is only slightly harder. but low gear is fairly low—2.20/1—and it really delivers a punch. These things happened with the dual-quad engine too, but never with the immediacy of the injected Chev, especially at low revs and low manifold vacuum.

The cold figures, available for your inspection, are pretty phenomenal. Injected and carb-fed Corvettes are closely comparable in performance, and both qualify as the fastest-accelerating genuine production cars SCI has ever tasted. In fact, up to 80 they're not so far from the data posted by the Mercedes 300SLR coupe, which is generally regarded as the world's fastest road car. In low, the Corvettes zoomed up to 55 mph in a shade over five seconds, and in another nine they surged to 95 in second with very exciting verve.

During the runs with the injected car, we were looking mainly for flat spots—the transition points in fuel/air metering that are often among the defects of the carb-fed, rather than injected, engine. There were none. The beef in the FI engine permits you to take off from standstill in top gear just by revving up to 1500 or so and letting the clutch out slowly. Our zero to 60 (actual) time *in top gear alone* was 13.8 seconds. Then we lugged the engine down to its smooth-running minimum in top gear, 14 mph, and accelerated using both full and partial throttle openings. There was no faltering, there were no flat spots. Power was generated smoothly and at an increasing rate until about 2800 rpm when the cam hit its stride and the car hunkered down and steamed into the distance. Elapsed time for 14 to 60 mph in high was 12.3 seconds, on the injected machine.

ABOVE: Chevy "Ramjet" constant flow injector is self contained, and sits in place of carbs. Better fuel distribution of FI allowed higher compression—10.5/1 as opposed to 9.5/1 on standard model. BELOW: Weak point in Chev's handling is rear. Leaf springs cannot take high torque, and cause axle judder.

1957 CORVETTES

PERFORMANCE

TOP SPEED:	Injected	Dual-Quad
Two-way average	125.0 mph	122.5 mph
Fastest one-way run	126.8 mph	123.3 mph

ACCELERATION:		
From zero to		
30 mph	2.4	3.1
40 mph	3.4	3.8
50 mph	4.9	4.9
60 mph	6.6	6.8
70 mph	8.8	8.7
80 mph	10.7	10.7
90 mph	13.2	13.5
100 mph	18.2	17.6
Standing ¼ mile	14.2	15.0
Speed at end of quarter	93 mph	95 mph

SPEED RANGES IN GEARS:	Injected	Dual-Quad
I	0-55	0-56
II	11-95	10-95
III	14-Top	16-Top

SPEEDOMETER CORRECTION: Indicated	INJECTED Actual	DUAL-QUAD
30	27	29
40	37	38
50	46	48
60	56	57
70	66	67
80	76	77
90	86	86
100	95	96

FUEL CONSUMPTION:	Injected	Dual-Quad
Hard driving	12.0 mpg	10.5 mpg
Average driving	13.6 mpg	13.8 mpg
Steady speed — 40	18.2 mpg	(Not checked)
50	18.3 mpg	
60	17.0 mpg	
70	14.7 mpg	
80	14.5 mpg	

BRAKING EFFICIENCY — **DUAL-QUAD**

(10 successive emergency stops from 60 mph, just short of locking wheels):

1st stop 64	6th stop 60
2nd stop 68	7th stop 63
3rd stop 63	8th stop 57
4th stop 54	9th stop 53
5th stop 63	10th stop 53

SPECIFICATIONS

POWER UNIT:	Injected	Dual-Quad
Type	V-8	
Valve Arrangement	Overhead in-line, pushrods	
Bore & Stroke (Engl. & Met.)	3⅞ x 3 ins. (98.3 x 76 mm)	
Stroke/Bore Ratio	0.773/1	
Displacement (Eng. & Met.)	283 cu ins. (4640 cc)	
Compression Ratio	10.5/1	95/1
Carburetion by	Rochester injection	2 four-barrel
Max. bhp @ rpm	283 @ 6200	270 @ 6000
Max. Torque, lb-ft, @ rpm	290 @ 4400	285 @ 4200
Idle Speed	950 rpm	500 rpm

DRIVE TRAIN:	Injected	Dual-Quad
Transmission ratios Rev	2.20	2.20
I	2.20	2.20
II	1.32	1.32
III	1.00	1.00
Final drive ratio (test car)	3.7	3.55
Other available final drive ratio	3.27, 3.36, 4.11	
Axle torque taken by	Rear leaf springs	

CHASSIS:	
Wheelbase	102 ins.
Front Tread	56.7 ins.
Rear Tread	58.8 ins.
Suspension, front	Parallel wishbones, coil springs
Suspension, rear	Semi-elliptic leaf
Shock absorbers	Delco tubular
Steering type	Worm and ball bearing roller sector
Steering wheel turns L to L	3½
Turning diameter	39 feet
Brake type	Self-energizing, 11 inch drums
Brake lining area	158 sq. ins.
Tire size	6.70 x 15

GENERAL:	
Length	168.0 ins.
Width	70.5 ins.
Height	52.0 ins.
Weight, test car	Injected 2840 lbs., Dual-Quad 2960 lbs.
Weight distribution, F/R	54.1/45.9
Weight distribution, F/R with driver	53/47
Fuel capacity — U. S. gallons	16½

RATING FACTORS:	Injected	Dual-Quad
Bhp per cu. in.	1.00	0.955
Bhp per sq. in. piston area	3.01	2.87
Torque (lb-ft) per cu. in.	1.02	1.01
Pounds per bhp — test car	10.00	11.0
Piston speed @ 60 mph	1400 fpm	1340 fpm
Piston speed @ max bhp	3100 fpm	3000 fpm
Brake lining area per ton (test car)		107 sq. ins.

Corvette R.T.

The same continuous flow of power held good during hard cornering and braking. Under these conditions with the dual-quad car, the contents of the carb float bowls were slung away from their intended orifices and left the engine gasping for fuel. There was no trace of this with the FI Chev, which feels as though it could be driven upside down.

In all the recent prophecies concerning FI, we've been assured that fuel economy will be improved. In SCI's tests, the FI version registered as much as 15 percent better gas mileage—in spite of a lower gear ratio. Part of this gain is due to the FI job's 10.5 compression ratio. In spite of the high compression our test car's engine did not detonate even when lugging heavily in top gear.

In the U. S. far more interest is shown in cars' sheer performance than in their fuel economy and we therefore normally restrict our fuel consumption figures to averages obtained in overall running. But because of the novelty of FI and because of the fuel consumption claims made for it, we ran precise fuel checks at steady speeds, using a Donat Gauthier 1/10-gallon fuel-measuring burette. The surprisingly good results for so potent an engine are shown in the data table. They suggest the fuel economy that's possible if you should choose to drive one of these injected cars sedately . . . which is not the kind of use they're built for. More important, they indicate the improved operating economy that can be expected when FI does become sufficiently inexpensive to be fitted to normal touring cars.

To get more expert opinion than our own on Chev's FI, we turned our test car over to racing FI specialist Stuart Hilborn, asking that he drive it and give frank comment. This was Hilborn's first behind-the-wheel encounter with Chev FI and he gave the car a careful, critical workout. Then he said, "It's good. All I can find to criticize is its hot-start behavior and its complexity."

As an addendum to the performance figures obtained, and to the comparison curve from our '56 Corvette test, we might mention the net horsepower ratings of the two 1957 cars tested—in other words, the actual output at the clutch with all accessory leads subtracted. The injected engine delivers 240 bhp at 5600 rpm, while the dual-quad rig turns up 230 bhp at 6000. Last year's Corvette engine was very modestly rated at 225 horses, so the improvement this year is just what you'd expect from an added 18 cubic inches.

Through a lot of flogging the Corvette clutch engaged clean and hard, yet was smooth enough to keep the car reasonably tractable in spite of its high low-gear ratio. There's nothing silent about the cogs in this three-speed Chevy box, and second emits a fine, healthy whine which rises to a high-pitched scream. Low has a growling howl, and in that gear the floor-mounted shift stick has an annoying chatter. Shifting is very fast, especially from second to top and back again. Powerful synchro eases that downshift, while the responsiveness of the injected version means no strain when double-clutching into the useful first gear. Chromed all over, the shift lever has a slippery feel. One test car had some sharp open threads just below the knob.

Last year we beefed mildly about the Corvette's behavior at a standing start, and in the meantime the horses have gone up while the rear suspension stayed the same. Now the rear end gets utterly out of hand when the clutch goes quickly home. The outboard-mounted leaf springs just can't take it, and the whole axle shudders and bounces. It took a nice balance between engine revs and deliberate clutch slippage to provide a smooth yet catapulting getaway. Leaving some form of radius rods or Traction Masters off this car is like forgetting the shock absorbers on a Cadillac. Installation of Chev's "Positraction" limited-slip differential is no direct solution either, as it would only make it tougher for the right rear wheel to break traction cleanly without jouncing.

The rear end is also the main weak point in the Corvette's handling. Up to the point of breakaway the front wheels are slipping somewhat more, due to the forward weight bias and the front end geometry, so the car is basically a strong understeerer. As the roll increases in a corner, though, which it does pretty rapidly, the high gear roll center builds up a heavy loading on the outside rear tire. At this point all that's needed to bust that wheel loose is a little more roll, a little tighter corner, a bit more speed or a bumpy surface. Any of these will ultimately cause rear end breakaway. At low speeds on a twisty bumpy road the rear gives ample warning for correction, but it goes early and combined with the understeer and the sheer size the Corvette can be a handful. On faster and better roads it's more at home, aiming and tracking very obediently, but on, say, damp concrete in a 75 mph bend the back end can be most unruly.

Steering is just quick enough to get out of most troubles but the strong caster action means you have to work to do it. This is responsiveness to muscle, not will. Emergency-type cuts are hard to take because the slim-rimmed wheel is placed very close to the driver's chest, and his arms get crossed up fast. The wheel itself is impressive with deep finger notches and drilled spokes, but its appearance is out of scale with the massiveness of the rest of the Corvette.

Generally, in fact, for a good-sized car the Corvette doesn't have very much room inside. The dog-leg at the windshield corner and the close-up wheel make it hard to get in and out, and the doors don't close solidly though they do have a strong hold-open catch.

To the eye the separately-adjustable seats look well shaped, but you still sit *on* instead of *in* them. Cushioning is firm and covering very smooth. Range of adjustment isn't wide; all the way back there's just barely enough leg room for the medium to tall driver. There's no excess room for the left elbow either. Head room is better in the convertible than in the hard top edition. In both cases visibility is good, the margin going to the hard top, and the curved windshield was distortion-free.

The pedals are easy to distinguish, heel-and-toeing the brake and throttle being possible. Though the headlight dimmer and windshield washer pedal are hard to tell apart, the latter works very well,

You have to stand hard on the brakes to get what stopping they can provide. Fade and low pedal came quickly both in our standard test and in hard mountain driving. One car suffered from unpredictable pulling, while the other delivered straight smooth (but not fast) stops every time. The Corvette rear suspension is annoying but since you're not constantly clutching off it's not a steady hindrance. We can't say the same for the brakes, which are an omnipresent menace in this otherwise capable road car.

A big red light just under the speedo gives warning when the pull-out hand-brake is on. Competition versions of the Corvette use this space for the tach, which would be an excellent production idea. This instrument panel was designed in 1952, and while it was never legible it has now also lost most of its aesthetic appeal. You can see the speedometer, but the graduations are hard to distinguish, and the rest of the dials are useless while moving. They're lit well enough, without windshield glare. Twisting the light control to the left turns on the two under-dash courtesy lights—handy for trips and rallies.

A check of riding qualities showed up another rear end fault: A tendency to bottom solidly on more than a few bumps. This was very annoying and is again the fault of the springs. We ran pressure in our test car tires on the high side of thirty, for a variety of reasons all connected with our personal welfare, and the resulting ride was definitely jouncy on rough surfaces. It wasn't too much softer with lower pressures. This tended to bring out rattles on both cars during the tests, but none were directly attributable to the Fiberglas construction.

Last year the power and chassis of the Corvette were well balanced, at last, and the car was good all around. For '57 it's faster, but the chassis is no longer fully up to the job. We hope that some of the research going into the Sebring SSR project will be diverted into production channels. Our tests show that Chevrolet Injection has a general margin over dual-quad carburetion, but this, frankly, isn't saying much. The chassis is just at the end of its string, while the engine and injection have a great future ahead of them. The 1957 Corvette may be the link between two eras.

Griff Borgeson & Karl Ludvigsen

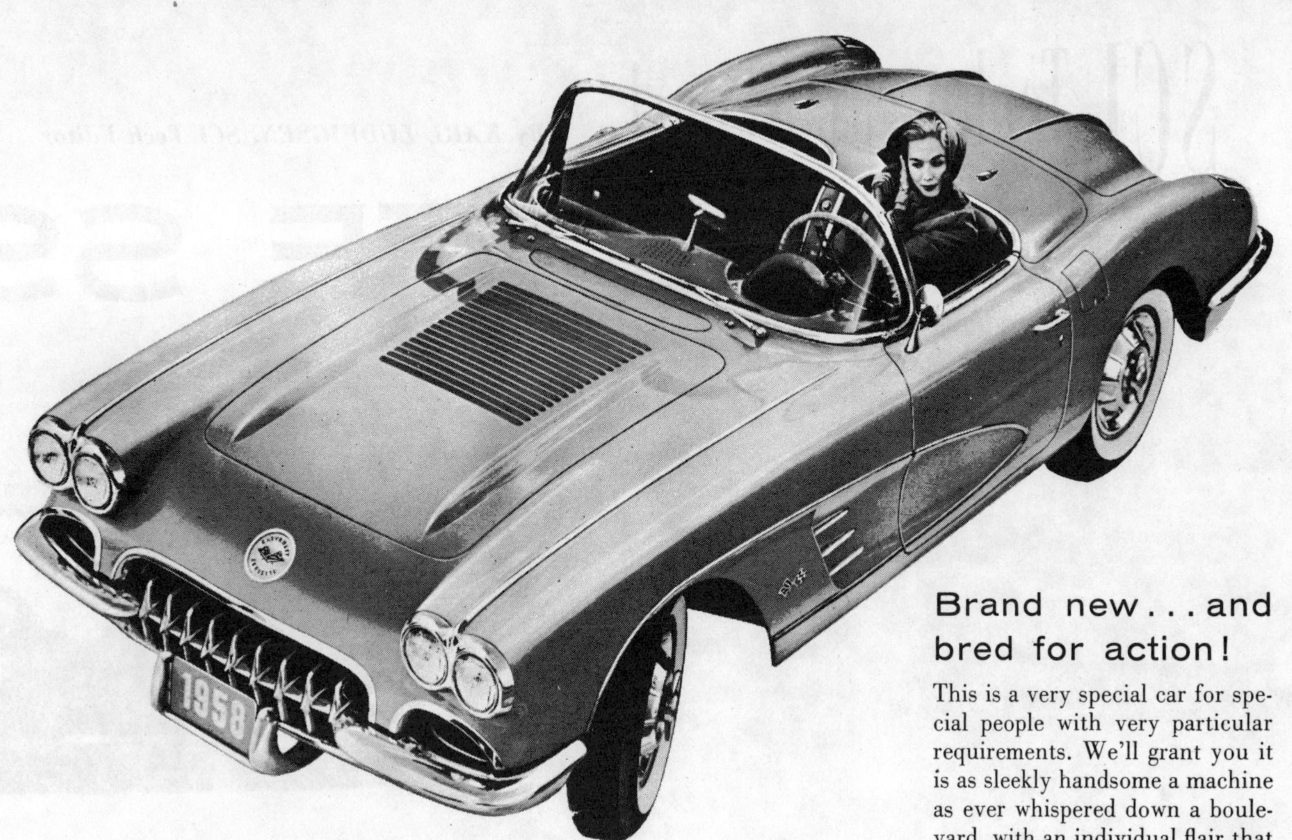
1958

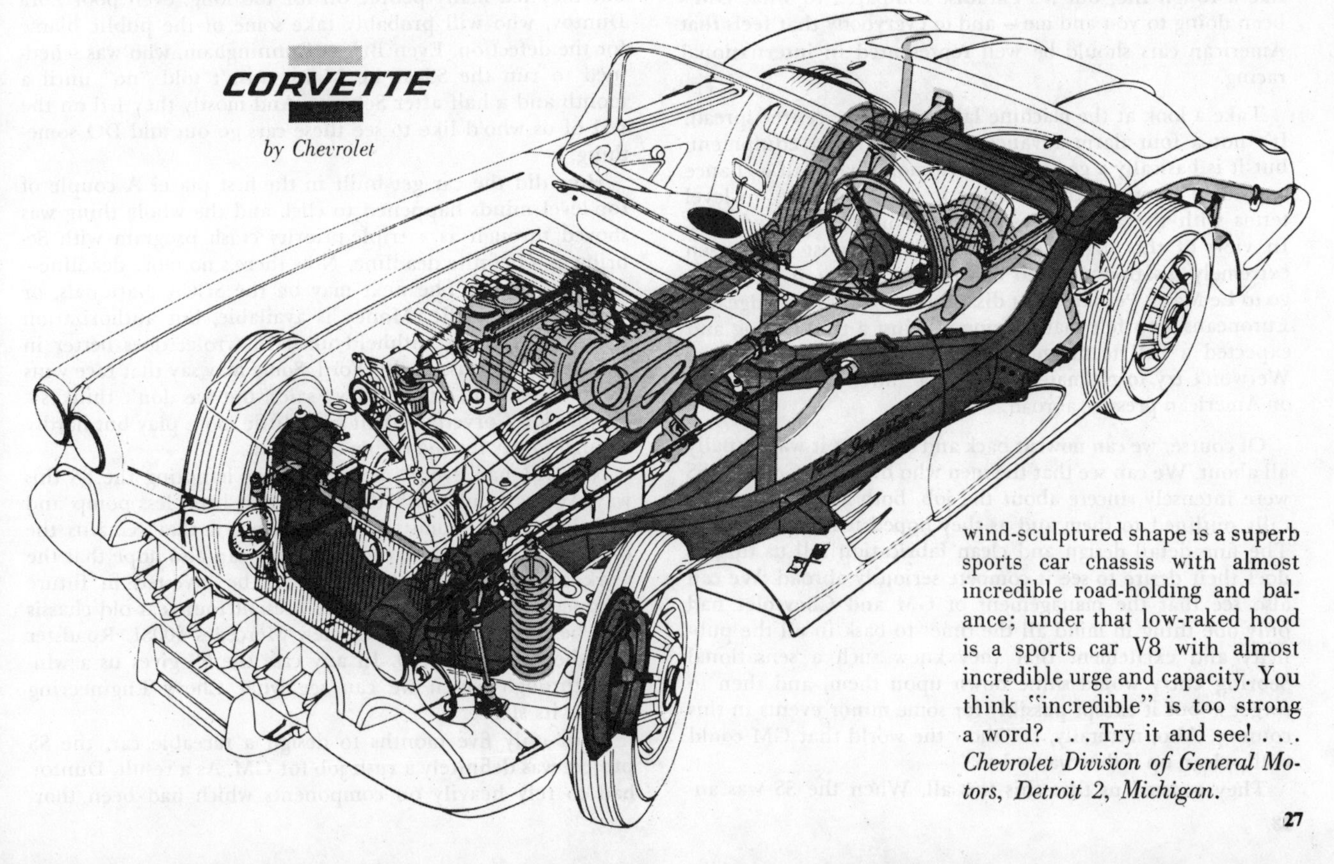

SCI Technical Report:

By KARL LUDVIGSEN, SCI Tech Editor

CORVETTE SS

REMEMBER the fable of Tantalus? This unfortunate gent was doomed to stand in the midst of a sea with clear, cool water right up to his neck and boughs laden with succulent fruits hovering over his head. When he bent down to drink the sea rushed temptingly away, and the boughs always swung just beyond his reach. This sounds like a rough life, but it's Paradise compared to what GM's been doing to you and me – and to everybody that feels that American cars should be well represented in international racing.

Take a look at the machine laid out on the center spread. It's not a four-alarm advance over all existing equipment, but it is basically a good car. Given more than half a chance and some intensive track testing it could compete on level terms with the world's best–Sebring practice proved that. In view of this highly publicized fact it must have been extremely disappointing to many to hear that it wasn't to go to Le Mans. Perhaps most disappointing to knowledgeable Europeans who felt that Sebring was just a trial outing and expected a full team and all-out effort for the 24-Hours. We won't try to estimate the effect a "no-show" could have on American prestige abroad.

Of course, we can now sit back and see what it was actually all about. We can see that the men who built the Corvette SS were intensely sincere about the job, both as it was specifically outlined to them and as they hoped it might develop. The fine detail design and clean fabrication tell us this, as does their desire to see it compete seriously abroad. We can also see that the management of GM and Chevrolet had only one thing in mind all the time: to bask in all the publicity and excitement that they knew such a sensational Sebring entry would shine down upon them, and then to forget about it except possibly for some minor events in this country. Also, naturally, to "show the world that GM could really clean up if they wanted to."

They warned us that this was all. When the SS was announced Chevy General Manager Ed Cole "emphasized that it is a research project to study advanced engineering characteristics in the field of performance, handling, braking and *other* safety features". The word we italicized is Chevy's loophole in case another Congressional committee shouts "Speedmonger!" This was all they did and do intend, but they led many people on for too long, even poor Zora Duntov, who will probably take some of the public blame for the defection. Even Briggs Cunningham, who was scheduled to run the SS at Le Mans, wasn't told "no" until a month and a half after Sebring. And mostly they led on the rest of us who'd like to see these cars go out and DO something.

How did the car get built in the first place? A couple of top-level minds happened to click and the whole thing was shoved through as a triple-priority crash program with Sebring as a definite deadline. Now there's no more deadline– no place to go. The next may be the SCCA Nationals, or perhaps Bonneville. Money is available, but authorization to use it is being withheld until Chevrolet does better in this little sales tussle with Ford. Some may say that race wins by the SS would boost Chevy sales, but we don't think so. Production Corvettes might get a little more play but hardly enough to pay for the racing operation.

No, GM had every justification for handling the SS this way. We can only wish that there had been less pomp and a little more circumstance at Sebring, if that remains the only major appearance of the SS. We can also hope that the work of Duntov and his crew will be invested in future production Corvettes, since the present four-year-old chassis may be pushed hard by the new Mercedes 300SL Roadster and the Jaguar XK150. In any case the SS gives us a window through which we can see what Chevy Engineering has up its sleeve.

With only five months to design a raceable car, the SS project was definitely a rush job for GM. As a result, Duntov had to rely heavily on components which had been thor-

oughly tested before, and could only lighten them if possible and fit a new framework around them. Fortunately a lot of miscellaneous information had been compiled from experimentation and racing with stock Corvettes and the special SR2 versions.

For one thing, they knew pretty well what the 283-inch V8 could and could not do. When the displacement was boosted from 265 there were some misgivings about the crankshaft, but undercutting the fillet radius at the journals has kept this glued together at 7000 and up. A weak point did show up at the wrist pin bosses in the piston, which distorted at high revs—notably in the badly overrevved SR2 at Nassau—and came apart. A little more meat around the boss cured that. Development on the SR2 for Daytona also led to the 40 inch tuned exhaust length that was incorporated in the SS. Racing during the winter helped to shake down the Rochester fuel injection system and determine its limitations.

Pressure of SCCA "Production" racing had forced the development of a four-speed gearbox, which with the use of an aluminum alloy case was just right for the SS. The iron case box, by the way, was available as of May first for $189 extra, or about the markup asked for the automatic transmission. Sounds encouraging.

The rush program for Sebring in 1956 turned up the sintered metallic and ceramic brake lining that's been used on most racing Corvettes with considerable success. They're fine if you don't mind replacing the drums fairly frequently and *warming up* the brakes before using them hard. A type of drum finning was also devised that appeared to give good results.

With these for a start Duntov had to build a light, compact car with handling of a very high order. Since time was short the 300SL frame was elected as a good pattern to follow, and the placement of the main SS chassis tubes resembles the SL very closely—NOT the D-Type Jaguar, as the rumors have run for so long. When the major members were set smaller tubes could be added for the particular requirements of this engine and suspension and to add stiffness where stress tests showed it to be needed. Big cross tubes connect the abutments for the front and rear coil springs, the rear mounts being nicely curved and drilled towers. Where parts like the brake servo cylinders are attached the frame tubes are square, to ease mounting, but otherwise they're round and about an inch in diameter. Particularly reminiscent of the Mercedes are the pyramided tubes at the cowl and the truss structure under the doors.

Front suspension resembles that of Chevy passenger cars in that the non-parallel wishbones are welded up of steel pressings, but the whole assembly is scaled down. Ball joints are fitted at the outer ends, and the wishbone frame pivots are rubber bushed. With more time metal-to-metal bushings might be installed for more precise control. A small-diameter anti-roll bar crosses the chassis under the suspension and is connected to the bottom wishbones by short links.

Nobody interested in fast cars will be shaken by de Dion rear suspension, but it is a novelty for Detroit machinery (except for notable show cars like the Le Sabre, Firebird I, the La Salles and Pontiac's Club de Mer, only two of which ran). Though the rear end looks confused, the curved one-piece de Dion tube is fabricated and located very neatly indeed. There are four tubular trailing arms, two of which are rubber-bushed to the frame just forward of each rear wheel. The upper arms angle slightly outward and are ball-jointed to the tops of the hubs. The lower arms however converge to the center of the axle tube and are fixed to the underside of the tube by ball joints at that point. A rigid yet light three-point location resulted, the frame mounting of differential and brakes relieving the tube and arms of drive and braking torque reactions.

The arrangement of the bottom trailing arms was one of the few brand-new features of the SS, but it's worth remembering that this was the source of one of the failures that

CONTINUED ON PAGE 33

CORVETTE SS — too little time, too many cooks, but . . .

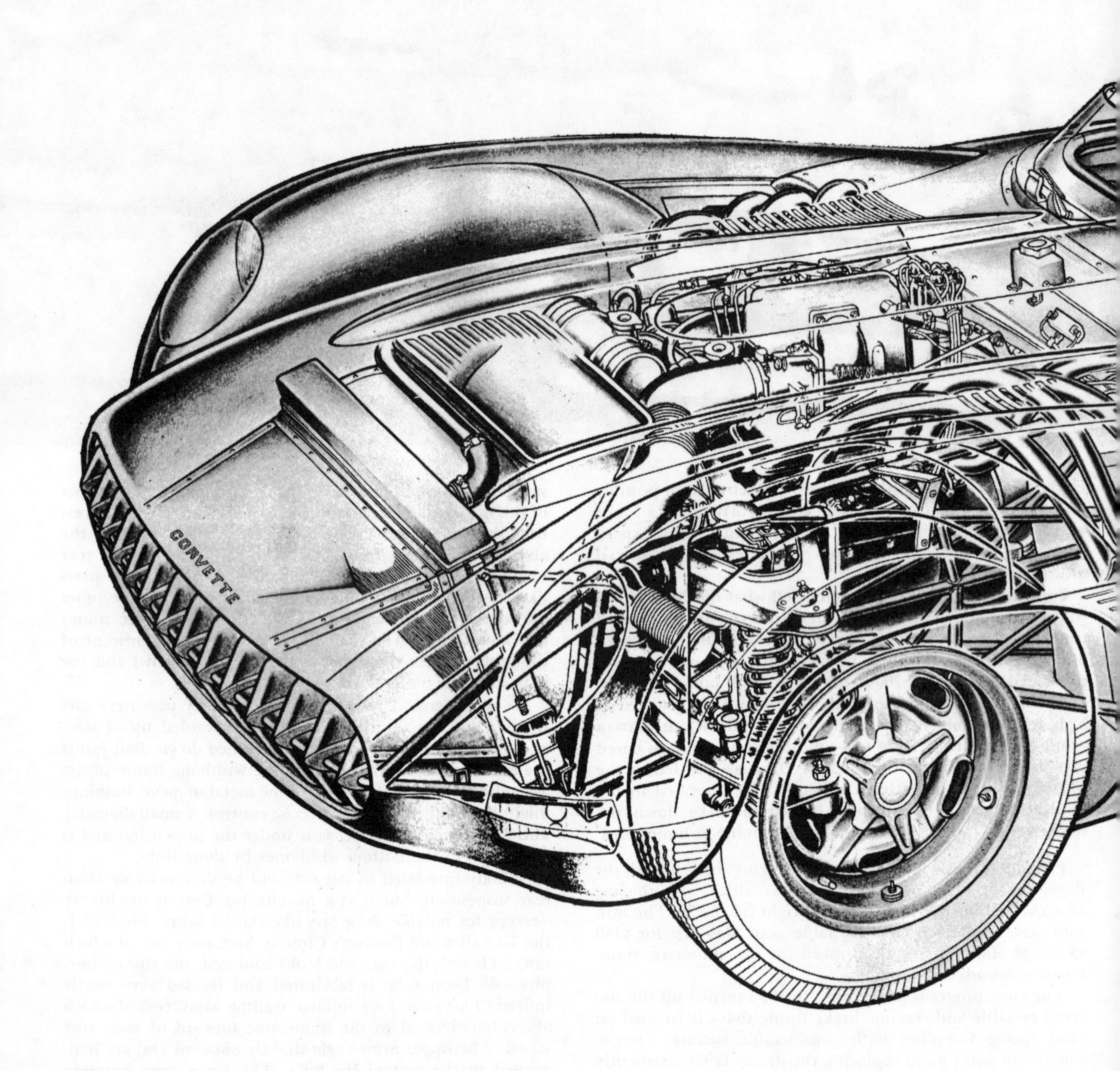

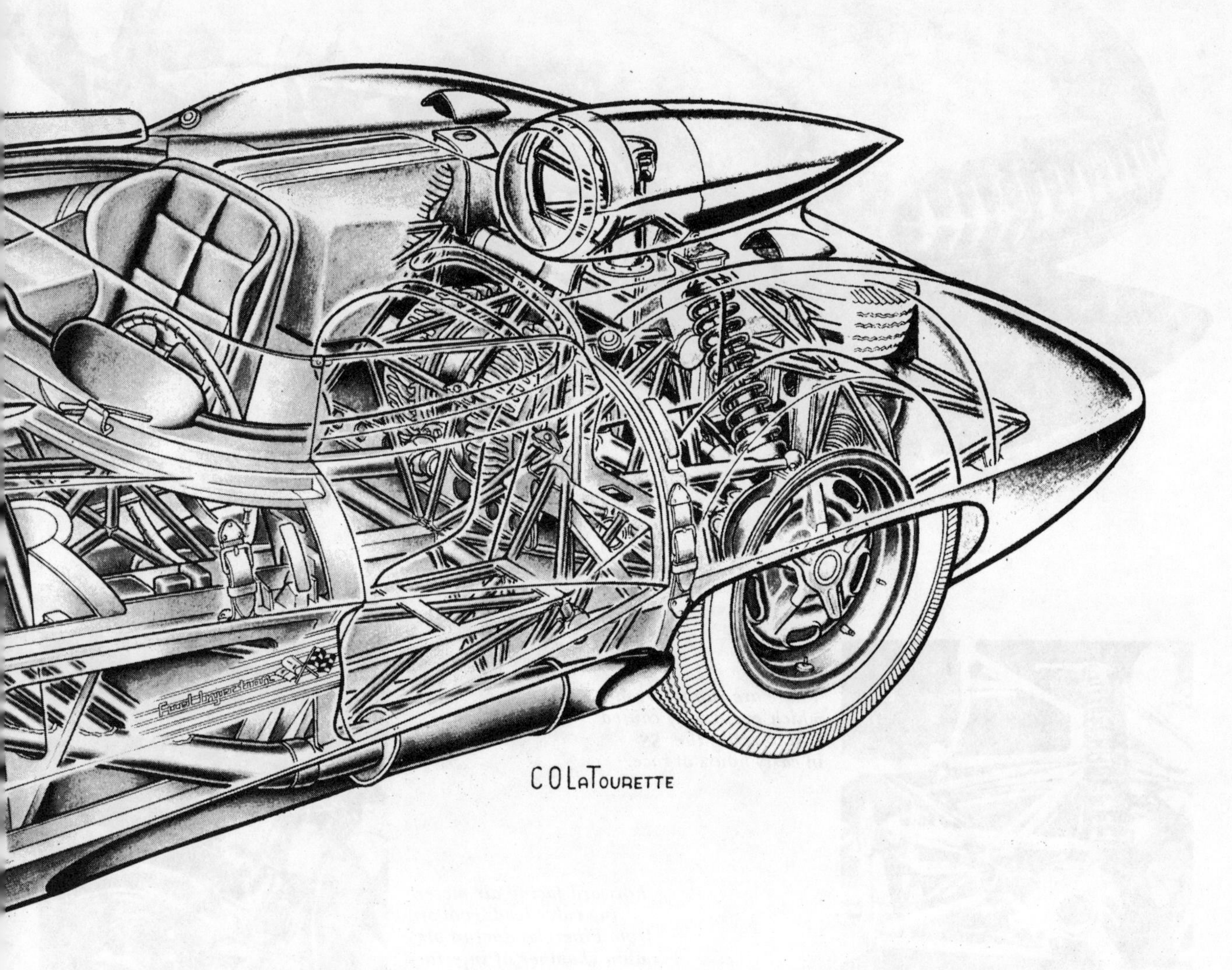

TEST CAR: CORVETTE SS

SPECIFICATIONS

POWER UNIT:

Type	V-8
Valve Arrangement	Overhead in-line, pushrods.
Bore & Stroke (Engl. & Met.)	3⅞" x 3" (98.4 x 76.2 mm)
Stroke/Bore Ratio	0.774/1
Displacement (Engl. & Met.)	283 cu. in. (4640 cc)
Compression Ratio	9/1 (11/1 optional)
Carburetion by	Rochester constant-flow fuel injection
Max. bbp @ rpm	310 @ 6000
Max. Torque @ rpm	295 @ 4400

Camshaft: Chain-driven "Duntov", 0.398" lift, solid tappets

Valves:	Intake	Exhaust
Opens	35° BTDC	76° BBDC
Closes	72°ABDC	31° ATDC
Diameter	1.85"	1.625"

DRIVE TRAIN:

Transmission ratios

I	1.87
II	1.54
III	1.22
IV	1.00
Reverse	1.87
Final drive ratio	3.87/1 with Halibrand quick change, Positraction differential.

CHASSIS:

Overall length	168"
Wheelbase	92"
Front Tread	51½"
Rear Tread	51½"
Weight, dry	1850 lbs.
Suspension, front	Non-parallel wishbones, coil springs, anti-roll bar.
Suspension, rear	de Dion, coil springs, four radius rods.
Shock absorbers	tubular, in unit with coil springs.
Steering type	Saginaw recirculating ball, three piece track rod.
Steering ratio	12/1
Brake system	One master cylinder, two separate servo systems.
Brake mechanism	Chrysler Center-plane 2LS with 12" x 2½" drums and Cera-metallic linings.
Wheel	cast magnesium with knock-off hubs.
Tire size	6.50/6.70 x 15 front, 7.10/7.60 x 15 rear.
Fuel capacity	43 U.S. gallons.

RATING FACTORS:

Bhp per cu. in.	1.096
Bhp per sq. in. piston area	3.30
Torque (lb-ft) per cu. in.	1.04
Pounds per bhp—test car	6.6
Piston speed @ max bhp	3000 fpm

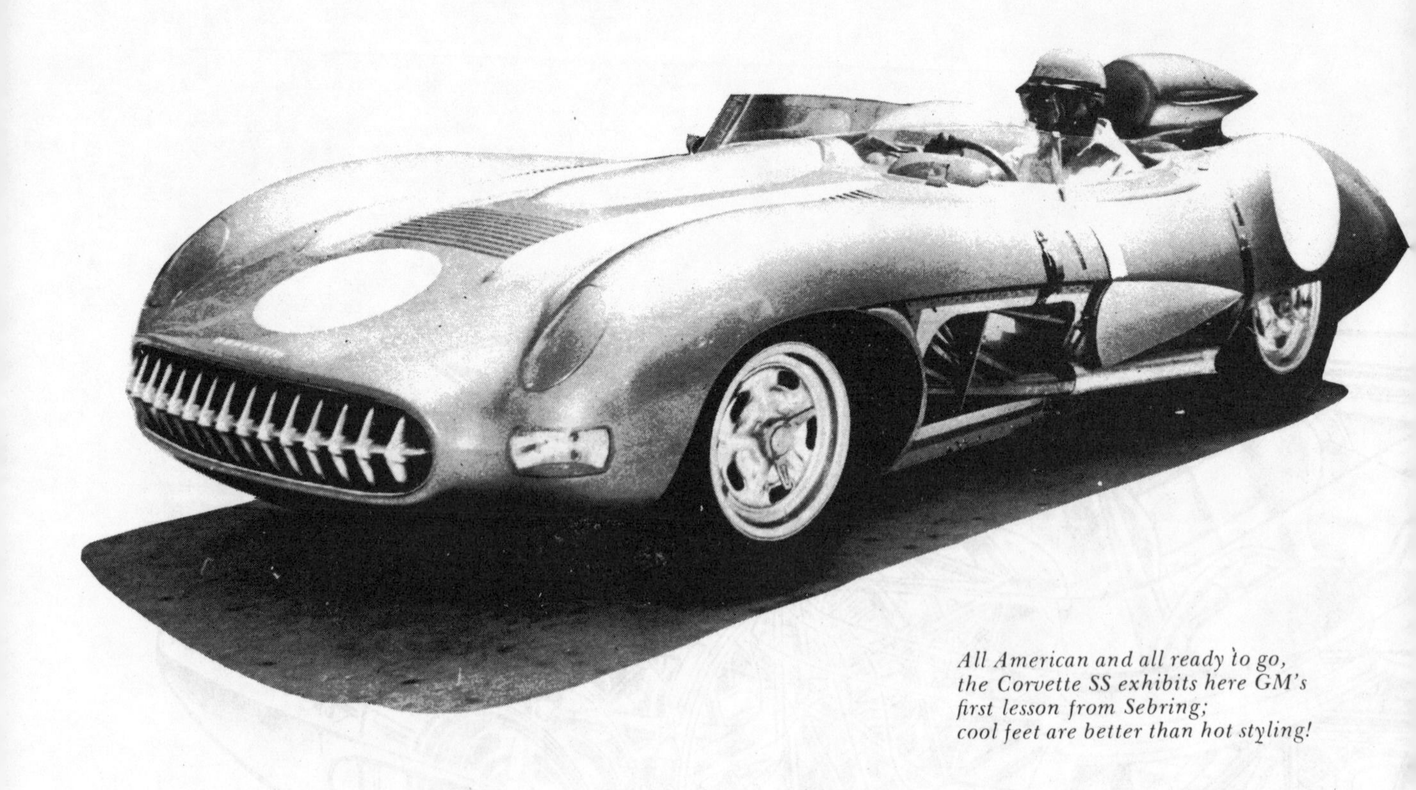

All American and all ready to go, the Corvette SS exhibits here GM's first lesson from Sebring; cool feet are better than hot styling!

There are rubber bushes which eventually caused GM to retire the SS in early hours of race.

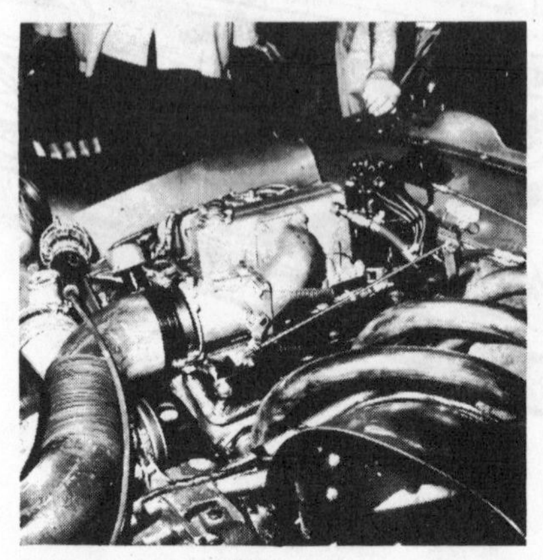

Forward facing air metering valve leads cool air from Fiberglas duct to plenum chamber of injector.

Everybody hustles to finish Chevrolet's activated dream car in time for debut at the Florida Grand Prix.

Details of finish are attended to with concern that might embarrass some Concours d'Elegance entrants.

Corvette

(Continued from page 29)

retired the car at Sebring. Rubber bushings are suspect in a suspension anyway, if very good steering is the goal, and one of them here did shift and destroy the alignment. Since this didn't show up on the much-flogged "Mule" it could well have been a material fault.

Springing at all four corners is by coil-shock units. The long small-diameter coils are carried in cups attached to the body of the tubular shock and its piston rod, giving a quickly demountable unit with a built-in bump stop. Rebound is limited by fabric straps. At first there was an additional housing around the coil, but this was tossed out to cut weight and allow quick access.

Probably the highlight of engine development on the SS was the use of aluminum cylinder heads for the basically stock 283 inch engine. These heads are very similar in design to the stock part, with only a slight repositioning of the intake ports to take advantage of some Weslake gas flow theories. They are definitely designed and run *without valve seat inserts.* Using the stock valve spring pressure of 210 pounds open and the slightly tuliped valves of the SS, pounding-in of the seats was very slightly more than normal, but not enough to cause any concern at all. If this technique can be reproduced, it could open up a brand new field in special heads for OHV engines. Only major structural change to accommodate the heads is the use of necked-down studs to compensate for the greater expansion of aluminum.

At Sebring it seemed that cooling troubles could be blamed on poor head gasket sealing, but it now looks like a subcontractor was to blame. Construction of the remote-mounted radiator header tank was farmed out, and a flow-control baffle was so misplaced that it cut off two thirds of the planned circulation. The tanks were peeled open and the baffles put in right. After that the ducted aluminum radiator performed as expected, as did the oil cooler incorporated in its base.

To the left of the radiator a Fiberglas duct scooped cool air into the Rochester injector machinery. The big air metering valve was faced forward instead of sideways to simplify the ducts and throttle control as much as possible. This injector requires a small air bleed to each nozzle for vaporization and idling, which is usually supplied by small pipes from the air cleaner. In this case there's a tiny individual filter for each adjacent pair of nozzles.

More important, nozzles for Chevrolet injection are now being built by the Diesel Equipment Division of GM, and are improved in two ways. First, the all-important nozzle size is determined by a thin calibrated disc instead of a lengthy sized hole, giving benefits in accuracy of distribution (which is still not so good with this system as it might be). Nozzle jet size is now .0135 inch instead of .0110. Second, each nozzle now incorporates a filter screen in addition to that at the pump. This has just about eliminated the chance of stoppage.

We weren't alone in wondering about the flex cable drive to the injection pump, *(SCI June 1957)* but this has been reliable except when the pump begins to jam, in which case the drive goes out before major damage is done.

The clutch and four-speed box are regular Corvette units, with the exception of the alloy housings that we mentioned, and the drive shaft is open with two universals. A late-model Halibrand center section houses a straddle-mounted pinion, helical quick-change gears and a Chev "Positraction" differential. Torque goes from here to the wheels through open axles with Hooke-type universals and sliding splined joints.

Since the differential is hung solidly from the spring support cross tube, it's tempting to mount the brakes inboard too and reduce unsprung weight. Duntov succumbed to this, as have many other designers, but Aston-Martin gave up this layout in 1953 for the good reason that heat from the brakes gets the differential hot, and vice-versa. It's not surprising that the same trouble is cropping up with the SS, but it can probably be licked by much better air venting down there. Additional scoops at the front end duct cooling air down into funnels attached to the backing plates.

As we've mentioned it was expedient to use two-leading-shoe Chrysler Center-Plane mechanisms to get brakes of the proper size and type in a hurry. GM devised their own drum design, though, which has also turned up on another Corporation product. The drums have a cast iron face and working internal surface, plus an aluminum finned muff which is locked mechanically to the outside of the working surface. 120 small holes are punched through the periphery of each drum, and when aluminum is cast around this it fills the holes and becomes a mechanical part of the cast iron. You can see also that the resulting internal drum surface will be dotted with little aluminum spots which can carry heat right out to the fins without passing through the iron at all! It's simple — almost crude — but it seems to work. Biggest danger is possible heat spotting from insufficient drum stiffness and uneven expansion.

These drums are 12 inches in diameter and 2½ inches wide. It was just recently announced that the Buick "75" for 1957 will be equipped with aluminum-finned drums of exactly this construction on the front wheels only, which have just the same drum dimensions. There's no reason to believe that it isn't the same part. So, if you'd like SS Corvette brake drums for your Chrysler, De Soto or '56 D500, go bang on the door of your Buick dealer!

Also interesting is the front/rear brake proportioning device used on all the Sebring Corvettes. This depends on two vacuum servo cylinders, mounted in the rear with the lightweight battery for convenience. The simple cowl-mounted master cylinder has a direct hydraulic connection to the right-hand servo cylinder, which in turn power-brakes the two front wheels directly. This requires two chassis-length hydraulic lines which are coil-wrapped for protection. In any case, then, the front wheel braking will always be proportional to pedal pressure, and will still be there if the vacuum fails.

Now, the hydraulic output of the left-hand servo cylinder is piped straight to the rear brakes, but the vacuum section is so linked to that for the front wheels that the two operate sympathetically. In other words, front brake force is directly controlled by the pedal, and rear braking is proportional to that at the front due to an air link between the respective vacuum cylinders. The basic front/rear proportion for the SS was set at 70/30.

Okay so far. The air pipe that connects the two vacuum cylinders can be sealed off by actuating an electric valve—undoubtedly a solenoid—which leaves the cylinder for the rear wheels completely isolated in whatever position it was when the valve closed. The electrical impulse is released by a mercury switch, mounted in the cockpit where it's handy. This switch is angled forward so that the mercury will slide up toward the end a given distance for a given quantity of car deceleration. When the mercury hits the end, in a stop of a preset negative "g", the solenoid is closed and rear braking force stays just as it was then— it can't increase; it doesn't go down. Front wheel force can then continue to rise in proportion to pedal pressure, but it's absolutely impossible to lock up the rear wheels no matter how hard you try! They're isolated from the circuit until the mercury switch and the valve open up again. With that mercury switch at just the right angle, braking at all four wheels can be fully used much more often than at present, when ultimate deceleration is limited by rear wheel locking. The switch angle could also be changed during a race to compensate for wet roads or different surfaces, or changing fuel loads. We've ridden in a car equipped with this rig, and think it's very promising.

If it gives more devices like these a tryout, the SS Corvette will be well justified as a rolling test-bed. Testing at Sebring before, during and after the race turned up a few basic faults in the SS' layout that are now being reworked. One, of course, was the extreme heat in the driver's compartment. To relieve this the front and rear pipes in the exhaust manifolds are being brought closer to the center pair, to pull the rear pipe away from the firewall. The pipes also curve more quickly to the outside of the magnesium body.

This manifolding, by the way, gave a big boost in power. The compression ratio on a competition version of the Chev V-8 had to be taken to 11/1 to get 310 horses, while the SS delivers the same amount on a 9/1 ratio with this exhaust.

On the bright side, there now exist three SS tube frames in addition to the "Mule" and the race car—a total of five possible machines. There are still a lot of guys at Chevrolet that believe in the SS and what it can do, and there's always a chance that Chevys will sell better in '58! Let's hope so, because this one is too good to be sent to the showers so early in the game.

—Karl Ludvigsen

SCI ROAD TEST:

THE '58 CORVETTE

TO mark the fourth birthday of the Corvette, its proud parents, the Chevrolet Motor Division, have announced the 1958 model which has undergone some extensive but not too important changes on the surface and a few rather interesting ones underneath. Starting right at the plastic body, the use of aluminum reinforcements in the cowl structure, inaugurated in mid-'57, has been extended to include the so-called "rocker panels" under the door openings. Bumpers are now bracketed to the frame in conventional American style, relieving the front and rear body panels of loads that are not rightfully theirs. These two items raise the weight "less than 100 pounds", but for racing, most of it can be unbolted and left in the pits without the SCCA batting an eye.

Uncowled dual headlights show how attractive most American front ends would be if we'd get off this "I'm longer than you are" kick. Just below them are really large holes for blasting fresh air onto the brakes, but on our test car, alas, the "holes" were painted black! More on this later on.

Further production experience with the F.I. nozzles and metering controls permits closer control over the air-fuel ratio this year. The warm-up diaphragm is now more sensitive and the air filter is also changed. On all Corvettes, the generator is now on the right-hand side so that the fan-belt engages the water pump pulley over a far greater arc, reducing slippage at high revs. Common to all '58 Chevy's with the 283 cubic incher are a new distributor rotor and a cap with longer sides to help keep out moisture.

Like most manufacturers, Chevrolet is none too happy about some of the attempts made to bring "boulevard" engines up to all-out F.I. specs. More is required than just a

The cornering of the "boulevard" Corvette cannot be described as flat, but to the driver it certainly feels very secure.

Duntov high-lift cam and a handful of solid lifters, although the factory is not too specific as to what is. What they have done is clarify the picture of available options.

First of all, here is what an absolutely standard Corvette would have (later we will get into what else can be ordered on the car at the time of purchase): The 283 cubic inch V-8 with a normal camshaft and hydraulic tappets (limiting revs to about 5500, as on our test car), a single four barrel Carter carburetor (#3744925), the "close-ratio" three-speed transmission (also used on other Chevy's with the 283 inch engine), a 3.70/1 ring and pinion, 6.70x15 tires (tubeless

Centrally located, the tachometer may now be readily observed, though it suffers from unwanted reflections off the curved lens, as do the other instruments.

The trunk space shows the American influence on sports car design; observes the Technical Editor, it's huge.

The louvers in the hood aren't real but everything underneath it is, and in a very big way.

Fitted with prototype linings, the brakes stood up to SCI's severe brake test very well indeed, considering that full-size wheel discs were worn during the test.

Accessibility of the fuel injection "box of tricks" is really great. There are lots of little bits and pieces but unlike carbs, here they're on the outside.

or not, to choice) on 5Kx15 disc wheels, and a choice of either the hardtop or the hand-operated folding one.

Options available that do not change the basic car's essentially boulevard character include: Powerglide transmission and with it, a 3.55 rear end; electric window equipment (which is no lighter than the hand-operated kind as reported elsewhere); a hydraulic mechanism for the folding top; and for the belt-and-suspenders types, both the hard and soft tops may be ordered on one car.

To improve performance, one can order either two Carter quads or fuel injection (we had the latter), the manifolds differing slightly between Powerglide and stick-shift cars. But for the most in "go", there is the 290 bhp @ 6200 rpm "D" fuel injection engine which features a 10.5/1 compression ratio, the high-lift cam, solid lifters, an air intake extension to bring in cool outside air, a reputedly "more efficient radiator", and a tachometer reading up to 8000 rpm. Especially designed for this engine, but definitely available on its own, as on our test car, is a really delightful, all-synchronized, four-speed gearbox.

In much the same category are the Positraction limited-slip differentials available with either the 3.70, 4.11, or the 4.56 ratios (and though you can't order it this way, the normal Chevy 3.89 gears will fit the carriers of either the 4.11 or the 4.56 Positraction diffs). To give slightly better side-load characteristics, wider (5½Kx15) rims are available for fitting 7.10 or 7.60x15 tires, racing or otherwise; the difference between the two enabling last minute "gear" swaps to be made at races.

For the guy who is really serious about his racing, a heavy-duty brake and suspension package is offered in an all or nothing deal. To get this package, you must also order the "D" engine and the Positraction differential. But what a package! Stiffer front coils give a spring rate 13½% higher. The anti-roll bar is 40% stiffer. The rear springs, with an extra leaf, have a 9½% higher rate. The shock absorbers, with 88% larger working area, have different valving and finally, the steering ratio is changed from 21/1 to 16.3/1 by lengthening the third arm idler.

The famous Cerametallic brakes are fitted and it is interesting to note that although the drum diameter remains at eleven inches and the shoes are a full half inch wider, the total braking area is actually reduced 20%, because the forward shoes are lined over only half their length. To reduce the amount of braking done by the rear wheels, the brake cylinders there are only 0.875 inch diameter instead of one inch, whereas the front ones remain at 1.125. The drums have cooling fins cast on the rim, and as a further option, vented backing plates with air scoops are available. Those large holes up front that we mentioned before may then be opened up and a duct will carry air back, not just to the front brakes, but under the door sills all the way to the rear ones, too.

The Cerametallic brakes are definitely not intended for all types of driving. Corvettes so equipped are delivered to the customer with a placard on the windshield which reads, "This car is not for street use". Until warmed up, they are quite apt to pull strongly to one side or the other; not just the thing for Grandma on her jaunts to the grocery store!

Faced with the realities of the American scene, Chevrolet now follows tradition in marketing two apparently similar, yet actually quite different sports cars, one for the every day sort of user who might occasionally go racing, and another for the serious competitor in the Production category. However, in this case, the engine mods from the racing model are readily available without the HD brake and suspension kit, which may seem rather the wrong way around. But at least you can't get the "D" engine with the

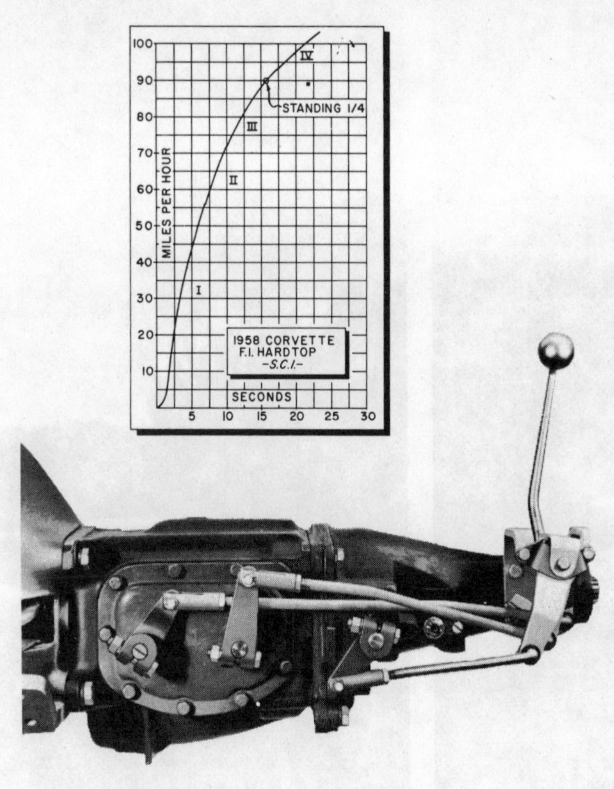

One of the Corvette's secrets of success is this really splendid 4-speed all-synchromesh gearbox.

Powerglide transmission! That *would* be too much!

One of the pleasanter aspects of this test was that, being in the nature of a sneak preview, the entire operation was conducted on GM's Proving Grounds at Warren, Michigan. After the brake fade and acceleration tests were completed on a 1½ mile level straight, we turned the Corvette loose on a sample road circuit that rather resembled Torrey Pines with its multiplicity of turns of varying radius, camber, and even surface texture. A visitor is said to have remarked naively that GM, with all its money, certainly could have afforded to build better roads than these. Be that as it may, we were able, in a very short time, to discover how the '58 Corvette behaves in nearly every conceivable road situation. Briefly, it may be summed up as "very well indeed."

There are no tricks at all to the steering, which is amazingly light at all times. We went through a series of ess-bends at speeds ranging from 40 to 70 mph. The only time the car felt at all uncertain was on a special piece of pavement featuring ridges running parallel to our direction of travel. The reaction here was pretty typical, the back end wanted to walk out somewhat when we crossed them on a diagonal. Elsewhere on the track, when we abruptly crested a sharp rise in the middle of a seventy miles an hour bend, the front of the car moved out only slightly, a tribute to a well-arranged front suspension and the high polar moment of inertia. On really tight hairpins, tighter ones than you have any right to be going that fast on, the steering is still light, though the steering lock seems to call for rubber arms. (The HD kit reduces the 3.7 turns lock to lock to under 3.)

Whether on fast bends or slow, when you reach the limits of adhesion, the back starts to come around in a calm, unhurried manner that leaves you plenty of time to get off

Continued on page **100**

1958 CHEVROLET CORVETTE F.I. HARDTOP

PERFORMANCE

TOP SPEED:

Est. 125 mph .... (see text)

ACCELERATION:

From zero to

30 mph	3.3 sec.	80 mph	12.2
40 mph	4.5	90 mph	15.7
50 mph	5.8	100 mph	21.4
60 mph	7.6	Standing ¼ mile	15.7
70 mph	9.5	Speed at end of quarter	90 mph

SPEED RANGES IN GEARS:

Corresponding to 750-5500 rpm

I	0-56 mph	III	13-93
II	10-72	IV	17-top

FUEL CONSUMPTION:

	Test Car	Competition
Racing	15 mpg	est. 8 mpg
Average driving (under 60 mph)	18.5 mpg	

BRAKING EFFICIENCY:

(12 successive emergency stops from 60 mph, just short of locking wheels):

1st stop	60	
2nd	60	
3rd	63	
4th	63	
5th	59	(rear wheel locked momentarily)
6th	63	
7th	63	
8th	60	
9th	59	(rear wheel locked momentarily)
10th	62	
11th	59	
12th	54	

SPECIFICATIONS

POWER UNIT:

Type	V-8
Valve Arrangement	Pushrod, in-line ohv
Bore & Stroke	3.875 x 3.00 in (98.4 x 76.2 mm)
Stroke/Bore Ratio	0.774/1
Displacement	283 cu in (4640 cc)
Compression Ratio	9.5/1 (10.5/1 with optional camshaft)
Carburetion by	Rochester constant flow fuel injection (one or two Carter quads optional)
Max. Power	250 bhp @ 5000 rpm
Max. Torque	305 lb-ft @ 3800 rpm
Idle Speed	750 rpm

DRIVE TRAIN:

Transmission ratios		Test Car	Optional
Stick shift:	I	2.20	2.21 (non-synchro)
	II	1.66	1.32
	III	1.31	1.00
	IV	1.00	—
	Rev.	2.25	2.21
Powerglide	Low	3.82-1.82	
	High	1.82-1.00	
	Rev.	1.82	

Final drive ratio (test car) .... 3.70 (hypoid)
Other available final drive ratios.. 3.55 (std for Powerglide), 4.11, 4.56
Limited slip "Positraction" differential available with the 3.70, 4.11, and 4.56 ratios.
Axle torque taken by .... Leaf springs

CHASSIS:

Wheelbase	102 in
Front Tread	57 in
Rear Tread	59 in
Suspension, front	Unitized, independent, unequal length wishbones, coil springs, 11/16" dia anti-roll bar (13/16" optional)
Suspension, rear	Semi-elliptic leaf springs
Shock absorbers	Tubular hydraulic, 1" piston diameter (1⅜" optional)
Steering type	Semi-reversible, recirculating ball, center-point linkage
Steering wheel turns L to L	3.7
Turning diameter	38½ ft right, 39 ft left
Brake lining area	157 sq in (121 sq in optional—see text)
Tire size	6.70 x 15 (7.10/7.60 x 15 optional)
Rim size	5K x 15 (5½K x 15 optional)

GENERAL:

Length	177 in
Width	73 in
Height	51 in
Ground clearance	6 in
Curb weight, factory data	2912 lbs
Weight distribution, F/R	52½/47½
Fuel capacity	16.4 U.S. gallons

RATING FACTORS:

	Test Car	Competition model with 4.11/1 gears and 7.10 x 15 tires
Bhp per cu in	0.88	1.02
Bhp per sq in piston area	2.65	3.07
Torque lb-ft per cu in	1.08	1.02
Pounds per bhp	11.6	est. 10.3
Piston speed @ 60 mph	1420 fpm	1550 fpm
Piston speed @ max bhp	2500 fpm	3100 fpm
Brake lining area per ton	108 sq in/ton	83 sq in/ton
Speed in IVth gear @ 1000 rpm	21.4 mph	19.4 mph

This comes under the heading of making photographs lie, or at least exaggerate in order to stress a point. The "boulevard" Corvette is shown just after cresting an abrupt rise while cornering near the limit. It really got "caught out." The heavy duty equipped model on the right, which certainly does corner more flatly, is shown at about the same place, but coming the other way. Long, flat turns are just its dish.

DUAL-PURPOSE SPORTS CARS FOR 1959

CORVETTE

ALONE amongst sports cars, Corvette sticks to the Detroit habit of annual changes. Though far from all-new, the '59 models incorporate some interesting innovations; so much so that we opted for a comparison test of old and new at GM's Milford, Michigan Proving Grounds. Though but two Corvettes were sampled, we came away with a headful of contrasting thoughts, for they were as different as night from day. The white one sported all the '59 body and interior changes on a strictly boulevard '58 chassis—normal suspension, three-speed box (we drove off in reverse at first) and the single-quad equipped 283 cu in V-8 engine. It was, in fact, a mobile display car for itinerant journalists. The other machine was a choice item indeed. Though camouflaged by the older body—washboard hood panel and chrome trim strips on the trunk—it had all the chassis mods for '59, which included a four-speed box, Posi-traction rear and a "works-built" full-house power plant. Just our meat, and no wonder; it's Zora Arkus-Duntov's own company car, a natural test bed for new ideas.

One of them fills a long-felt need. To better control the movement of the relatively heavy rigid rear axle, radius rods have been fitted to it above each rear spring. Running forward to the boxed frame, they are merely shortened versions of the Panhard rod which will be standard on '59 Chevy passenger cars. Thus relieved of a burdensome additional duty, the rear shock absorbers are now mounted straight up and down and are recalibrated. Result: a slightly softer ride and noticeably less rear-end steering on irregular surfaces.

Two changes permit still faster shifts than ever on the excellent all-synchro four-speed transmission. Clutch pedal travel, normally 6.4 in, may be reduced to 4.5 by rearranging part of the clutch linkage. And to avoid getting lost in the reverse gate when snapping off a downshift from third to second, the spring-loaded detent has given way to what the GM people call a "positive-action reverse inhibitor for 4-speed transmission." Translated into everyday mechanical lingo, this means a T-handled positive lock mounted on the shift lever which you pull up with two fingers as you shift the lever over with the palm of your hand into the reverse gate.

The above changes, all very nice ones, are all standard items. The biggest news and the best is going to start out as an option. This is in the brake department where Moraine sintered metallic lining pads will now be available on any Corvette, with or without the heavy duty suspension, and, we suspect, eventually on any Chevrolet at all. Like the Bendix cerametallic linings, these attack the brake fade problem from the fade resistance angle rather than from the heat dissipation point of view, as Buick does with their now-bonded aluminum cast iron drums. But Moraine engineers have achieved, after more than a year of testing by Mauri Rose, what could not be done with the cerametallic pads; namely, keeping the drums from wearing out before the pads, and avoiding brake-grab when the stoppers are still cold.

Another problem with the "Not for road use" Bendix binders was that once the drums were scored, which was generally pretty darn soon, the braking was apt to be erratic. It was just enough to make the Corvettes twitch a bit at the ends of the long straights. Mr. Duntov says that despite this roughness, these cerametallic set-ups could last through *two* Sebrings in a row. Indeed, while we were in his office, he received a call from an owner-driver on the Pacific Coast who has just finished his second consecutive season on the same set of drums and shoes.

On the H-D brake kit, thirty-six pounds are eliminated by knocking out the air duct to the rear brakes. The front scoop now directs air to the front brakes, which were somewhat blanketed by the body panels that were widened last year. "Elephant ears" now cope with such problems at the rear.

For testing purposes we were again allowed to run riot on the Ride Road. This is a twisting, narrow road that every last one of us would be delighted to have in his backyard. Our first run was in a much modified Impala; but more about that some other time.

Next sample was the '59 bodied, '58 chassis boulevard sports car style of Corvette, the white one in the photos. Briefly, even with the mild, mild engine, wheelspin was a real problem on smooth blacktop starts. One more frictionful surfaces, wheel hop replaced wheelspin. There's a wide gap between first and second gear ratios, but with such a mild engine, the Corvette pulls easily in second from very low revs. At the west end of our little race course is a bend of about 600 feet radius. It goes around and around, about 225° in fact, but at the end there are some rather mean dips and rises. Staying in second for this one, we found that the car rolls considerably and that occasionally it would "run out of gas." The float chamber was being tilted too much. When leaving the dip we came down to earth in a solid fashion. The suspension bottomed out with a thud, but remarkably the Corvette held to its course, though displaced sideways a bit.

The other car was Duntov's Gray Ghost, and an impressive machine it was. The clutch was set up for short travel, correspondingly the pedal forces are slightly increased. Combined with the more forceful response of the 290 hp mill to the throttle, the whole set-up has a much more purposeful, taut feel. We had only made two passes before a torrent of rain deluged us, but the improvement in handling was eye-opening. It was much more level when cornering, had quicker steering, and it had consistent response to the throttle no matter what the attitude of the car.

The latter is most important in a car this powerful, for once the back end has started to "hang out," much if not most of the steering is done with the throttle rather than the wheel. Even in the rain this H-D Corvette showed remarkable stability. Some credit should go to the Firestone SS170 tires, but only some. They grip exceedingly well up to a point, but when you press on still harder it's almost like driving on ice. These were not the latest 170-T's mentioned elsewhere in this issue, incidentally.

The sintered metallic brakes on this car were strikingly unobtrusive. It was only after a dozen laps of our "circuit" that the constant spray of water chilled them enough that they became a bit touchy. Strictly a temperature effect, as inorganic linings do not absorb water.

Despite the commendable lightness of control, the Corvette, even in racing trim, seems to be an awful lot of automobile to jockey around a narrow road circuit at racing speeds. Despite its shortcomings, it is, as we said last year, hard to beat on the basis of performance per dollar of original investment. Though tire wear will keep running costs up, the "extensive use" of American parts should more than offset this. And if you are buying a Corvette with the intention of racing it, it is naturally cheaper to get it with the full factory-installed kit of racing gear than to try to bring a lesser car up to racing specs yourself. *sfw*

Photos by Dick Spare and Ted Fredley

Even when the rain started pouring down, Duntov's Gray Ghost stuck firmly to the road. Perhaps it just knows its own way round the Proving Grounds.

External changes for '59 are strictly customizing-type work. Washboard on hood, streaks on trunk are gone. Good riddance.

SCI ROAD TEST

'59 CORVETTE

Banked track hides considerable roll as Zora Arkus-Duntov places Corvette close in.

WITH EACH ANNUAL change, Zora Arkus-Duntov, the Corvette's godfather, has emphasized performance improvements. His theory is that to sell, the Corvette must first go. Styling has had its innings, too, but they have acted with more restraint than one expects from Detroit.

Perhaps in acknowledgement to the discriminatory taste of the sports car market, external changes for '59 were of a customizing nature only. The washboard-like, phony louvers on the hood and the Pontiac-like silver streaks on the trunk lid are now things of the past. Inside, there are recontoured seat cushions, a reverse lock on the four-speed's shift lever and an open-at-the-top catch-all which fills the opening in front of the "sissy-bar". Also the door-knob and arm-rest have been moved. Engineering changes at the rear include newly added longitudinal radius rods to prevent axle wind-up and re-arranged and recalibrated shock absorbers.

In discussing the new seats with Mr. Duntov, he pointed out that this is one of the most difficult compromises to make in a high-performance car. "In a racing car, the seat is 100% for working, just like a stool by a lathe. But in a passenger car, you're lucky if it's a 'work chair' more than 10% of the time. It must be easy to get in and out of, and comfortable for lounging in. The high sides of a real bucket seat are just right for holding you in place but they don't meet these other requirements."

Opening the trunk, later, we discovered an experimental seat cushion. We tried it and found it lets you sit a lot lower, accentuating the side support of the cushion edges and increasing headroom, too. The secret is foam rubber in place of the usual coil springs, the drawback is that it bottoms out too quickly on large bumps — sooner than the suspension. Oh, well, back to the drawing board.

Commenting on the door handles, Mr. Duntov pointed out that after driving Corvettes for thousands of miles on the Proving Grounds, he decided to move them forward during a visit to Riverside, California when he had to borrow a raincoat (?). The one he got had cinch-straps on the sleeves and he kept opening the door when turning left!

The sissy bar is unfairly named, it's really quite the thing for hanging on when the driver is trying to prove something or other. The trash tray appeals to us, too, though others have condemned it as unsafe. We suspect that someone must have been road testing with their knees tucked under their chin because it just isn't that low.

The world's largest producer of automobiles, Chevrolet offers not just a wide range of models, but a staggering array of optional equipment on each model. As on the sedans, so on the Corvettes. Five variations on the 283 V8 theme are offered, three transmissions (two, three, and four speeds), four final-drive ratios, three brake lining choices and two suspensions, just to name the mechanical ones.

There is some interlocking involved, for instance, you can't order the Positraction limited slip differential with Powerglide; the latter must have the 3.55/1 rear end and cannot have the 290 hp version of the Fuel Injection engine. On the other hand, the stiff suspension is available only in combination with the latter engine and Positraction.

For this test, the editors of SCI checked through the list of equipment options, nominating those we thought would add up to the most desirable all-around Corvette that you can buy. We had in mind, not the all-out racing version sampled in the December issue but a happy compromise that would be suited to both serious traffic and casual racing.

Having carefully selected our list, we were pleasantly surprised to find that the Chevrolet Division could put such a sample at our disposal immediately. Seems a fellow named Arkus-Duntov has a company car fitted out identically but for one exception. His car also has the quick-steering adaptor (3.25 instead of 3.7 turns lock to lock). Generally available only as part of the heavy-duty suspension kit, it's worthwhile on its own if you can stand the noticeably stiffer steering. And if you can get it.

Omitted from our list were the heavy-duty brake and suspension kits. For 1959, the heavy duty suspension features much stiffer springs than last year. In pounds-per-inch, the spring rate on the standard suspension, last year's HD kit and this year's are, respectively, 300, 340, and 550 at the front and 115, 125 and 145 at the rear. Though the kit's anti-roll bar diameter remains at 13/16-inches, the vast increase in spring stiffness contributes to both much flatter turns and much harsher bouncing. Nice for racing only, but not for the all-around usage we have in mind.

Corvette cornering has been the butt of many rude remarks by the anti-roll brigade. Duntov had interesting thoughts on this matter too. "For flat, smooth courses, such as Le Mans or Sebring," he said, "the heavy duty suspension option is very effective. But because it is so much stiffer, especially in roll, it would actually be a hindrance on a bumpy circuit such as the Nürburgring." An interesting thought, and interesting examples, too.

With so much power so freely available, rapid cornering necessarily becomes a maneuver requiring careful control of all the elements involved.

Though a stiffer anti-roll bar would reduce the independency of the front end, the reduction in roll would certainly promote more driver security. The experimental seats helped, so did the seat belts.

The steering, stiff for parking, was fine for controlling incipient slides, but as before, we found the throttle linkage much too sensitive. As a result, the car again appears to have two personalities. Either you motor sedately (though deceptively

Re-shaped seats increase lateral support.

Left, sintered linings; right, Cerametalix.

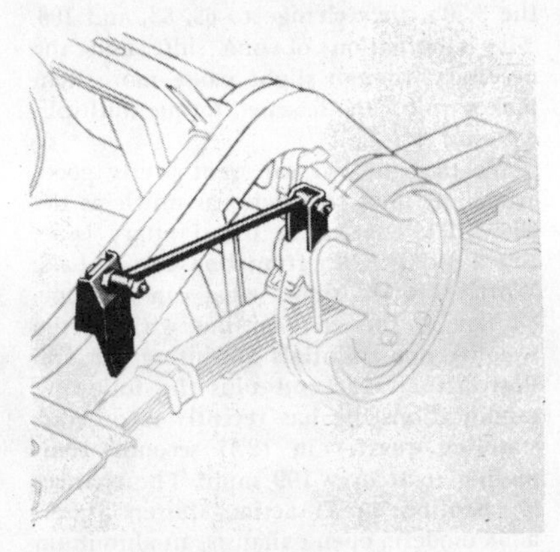

Axle wind-up prevented by radius rods.

quickly) through a corner, or pressing on somewhat, you're herding an untamed beast, one which responds more to the throttle than the wheel, and rather violently, at that.

In making up our list, the Cerametalix brake linings lost out as being entirely unsuited to highway use. When cold, they grab violently and erratically. Besides, they come only with the HD suspension.

Using regular drums, the sintered metallic brake linings at only $26.90 are so good and so cheap that they should be standard equipment. Though the 15° F. temperature during our test helped their heat dissipation, it gave us a chance to prove that they have no nasty habits when cold. Being inorganic, they don't soak up water either.

Flared drums, standard on all Chevies this year, have a bell rim-like flange which helps scoop air in from the inboard side. Between these and the linings, we repeatedly made stops from 90 and over without a trace of fade or increase in pedal travel. Pedal pressure, though, is higher than with organic (asbestos) linings. Running against a mild breeze, we managed zero to 100 mph and back to a standstill in 26.4 seconds, a figure that could be improved upon with experience.

Though it runs $484.20 more than the standard engine, we were anxious to see what the fuel-injected, hot cam V8's 290 horsepower felt like in a roadable Corvette. To its great credit, it must be mentioned that cornering antics have no effect whatsoever on throttle response. This is in great contrast to quad-equipped cars, whether single or dual. They will just plain quit in the middle of a hard corner.

Racer Brown's chart (see Part II of his article in next month's issue) indicates that the all-out dual-quad version boasts better output than FI-cars at the top end. However, the ability of the latter to mix correct quantities of fuel and air, no matter what the car's gyrations or speed, assures it of better lap times on any circuit.

That one Corvette is not the same as another is evident from the variety of options. What comes as a considerable surprise is the dualistic personality of the particular car tested. In the twinkling of a throttle linkage, it turns from a submissive, sidewalk stalker to a fierce, roaring eater-upper of metallic monsters. Just as quickly, it reverts to silent smoothness, its exhaust murmuring, barely audibly, "Who, me?"

This high output engine enjoys such tractability that it will pull smoothly from its 750 rpm warm idle in fourth gear. No flat spots, no hesitation, and no matter how fast you mash the throttle. Just to rub it in, you can start from rest in any gear whatsoever without stalling and with hardly a thought of the clutch lining. Strangely, starts in first gear seem equally touchy. Seems the "fast" clutch linkage (you have your choice of 4.5 or 6.4 inches of travel) has the unfortunate side characteristic that its mechanical advantage decreases quickly just as engagement occurs. For city traffic, we found that the combination of close ratios and the wide rev range made use of just first and fourth gears quite acceptable.

Despite this, the four speed gearbox ($188.30) was high on our list. For real get-up-and-go, it is well laid out. Shifting at 6500 rpm each time drops the revs only to 5000 or so, where the V8 is pulling strongly. To compare with the three speed box, all you do is skip second. Then coming out of first at 6500 drops revs to only 3900. Quite a difference when you're trying to make tracks.

Acceleration, while breaking no SCI records, is tops for pushrod equipment, and at the top end, quite similar to the 3.0 liter Ferrari 250 GT tested last year. Not so good off the line though, proving that smoothness at low revs is not synonymous with strength.

Duntov, in his kidding, quizzical way, said, "You know, the 3.70 axle gives better acceleration times than the 4.11." The operative word is "times", and specifically the times for zero to sixty, eighty and one hundred.

Assuming one is driving the car for best acceleration down a quarter-mile, (which is what we're after in our road test runs), this is quite true. With the 4:11 gears, shift points are 57, 75, and 95½. But with

Shapely body plays down large proportions of Corvette, derived from stock suspension.

the 3.70's, they change to 63, 83, and 106. The elimination of one shift and the necessary, though slight pause more than makes up for the lessened torque multiplication.

We thought our car went pretty good, even if it does cost twice as much as the Chev "315" tested in our January issue. But a recent letter from Don Gist of Lake Worth, Florida shows what can be done. Similar to this test car but without the weighty non-essentials and using the 4.56 Positraction rear end plus the following modifications, he has recently recorded a standing quarter in 12.57 seconds, completing it at over 109 mph! The changes he mentions are Traction Masters (his is a '58 model), open exhausts, an aluminum flywheel, Packard 440 wiring, 7.60 x 15 Bruce Slicks and the all-important flywheel shield. He adds, that the engine was "properly tuned." Uh-hunh.

If you want a particular kind of Corvette, pick out the options you want and GM'll make it. If they haven't the options you want, your local speed shop will.

If your interests lie in drag-racing, by all means pick the 4.56. If road-racing is more your style, then you'll probably want either the 3.70 or the 4.11, depending on which course you're running. Remember. 6500 in top corresponds to 139, 125, and 113 mph respectively with the 3.70, 4.11, 4.56 ratios.

For straight highway use, the 3.70, which we would now prefer, will provide the minimum of engine noise with plenty of acceleration. And if the chips are down, you can always use the gearbox. The 3.70 might improve gas mileage, but a lighter foot would help more.

On the luxury side, we chose the station seeking push-button radio ($149.80), the heater-defroster ($102.25), windshield washers ($16.15) and, sybarites to the end, both the folding canvas top and the removable hardtop ($236.75). Added to the $3875.00 base price, the total climbs to $5127.80; transportation and state and local tax are extra.

It's not a low price car, and it's none too cheap to operate, but it goes well, it stops well, and with reservations, it corners well, too. For all around performance per dollar, the Corvette is hard to beat.

—Stephen F. Wilder

'59 CHEVROLET CORVETTE

PERFORMANCE

TOP SPEED:

Estimated .... 125 mph

ACCELERATION:

From zero to	seconds
30 mph	3.2
40 mph	4.2
50 mph	5.2
60 mph	6.6
70 mph	7.9
80 mph	10.1
90 mph	12.3
100 mph	15.6
Standing ¼ mile	14.9
Speed at end of quarter	98 mph

SPEED RANGE IN GEARS:

(700-6500 rpm, 7000 permissible)

I	0-57 mph
II	8-75
III	11-96
IV	14-125

SPEEDOMETER CORRECTION:

Indicated Speed	Timed Speed	Indicated Speed	Timed Speed
30	29½	70	65
40	38	80	74
50	47	90	83
60	56	100	92

FUEL CONSUMPTION:

Hard driving .... 10 mpg

SPECIFICATIONS

POWER UNIT:

Chevrolet 283-FI	Water-cooled V-8
Valve Operation	Pushrods and stamped rockers
Bore & Stroke	3.875 x 3.00 in. (98.4 x 76.2 mm)
Stroke/Bore Ratio	0.774/1
Displacement	283 cu. in. (4640 cc)
Compression Ratio	10.5/1
Carburetion by	Rochester fuel injection
Max. Power	290 bhp @ 6200 rpm
Max. Torque	290 lbs.-ft. @ 4400 rpm
Idle Speed	750 rpm

STANDING 1/4

MILES PER HOUR

SECONDS

1959 CORVETTE 290 H.P. FUEL INJECTION —S.C.I.—

DRIVE TRAIN:

Transmission ratios	test car	optional ratio
I	2.20	(2.21)
II	1.66	(1.32)
III	1.31	(1.00)
IV	1.00	
Final drive ratio	4.11	(3.70, 4.56)
Axle torque taken by	rear springs and radius rods	

CHASSIS:

Frame	Welded box section side members, I-beam X-member, box section front and rear cross members
Wheelbase	102 in.
Tread, front and rear	57, 59 in.
Front Suspension	Unitized, independent, coil springs and unequal wishbones, 13/16" dia. anti-roll bar.
Rear Suspension	Rigid rear axle housing, semi-elliptic leaf springs, upper radius rods
Shock absorbers	Delco telescopic, 1⅜ in. piston diameter
Steering type	Saginaw worm and ball bearing sector, 16.3/1 ratio
Steering wheel turns L to L	3.25
Turning diameter, curb to curb	37 ft.
Brakes	Sintered metallic linings in composite drums with cast iron rim, pressed steel web
Brake lining area	108 sq. in.
Tire size	6.70 x 15
Rim size	5.0 x 15

GENERAL:

Length	177 in.
Width	73 in.
Height	51½ in.
Weight, as tested	3400 lbs.
curb (factory figure)	3092 lbs.
Weight distribution, F/R as tested	53/47
Fuel capacity	16.4 U.S. gallons

RATING FACTORS:

Specific Power Output	1.02 bhp/cu. in.
Power to Weight Ratio, laden	11.7 lbs./hp
Piston speed @ 60 mph	1560 ft./min.
Braking Area per ton, laden	63.6 sq. in./ton
Speed @ 1000 rpm in top gear	19.2 mph

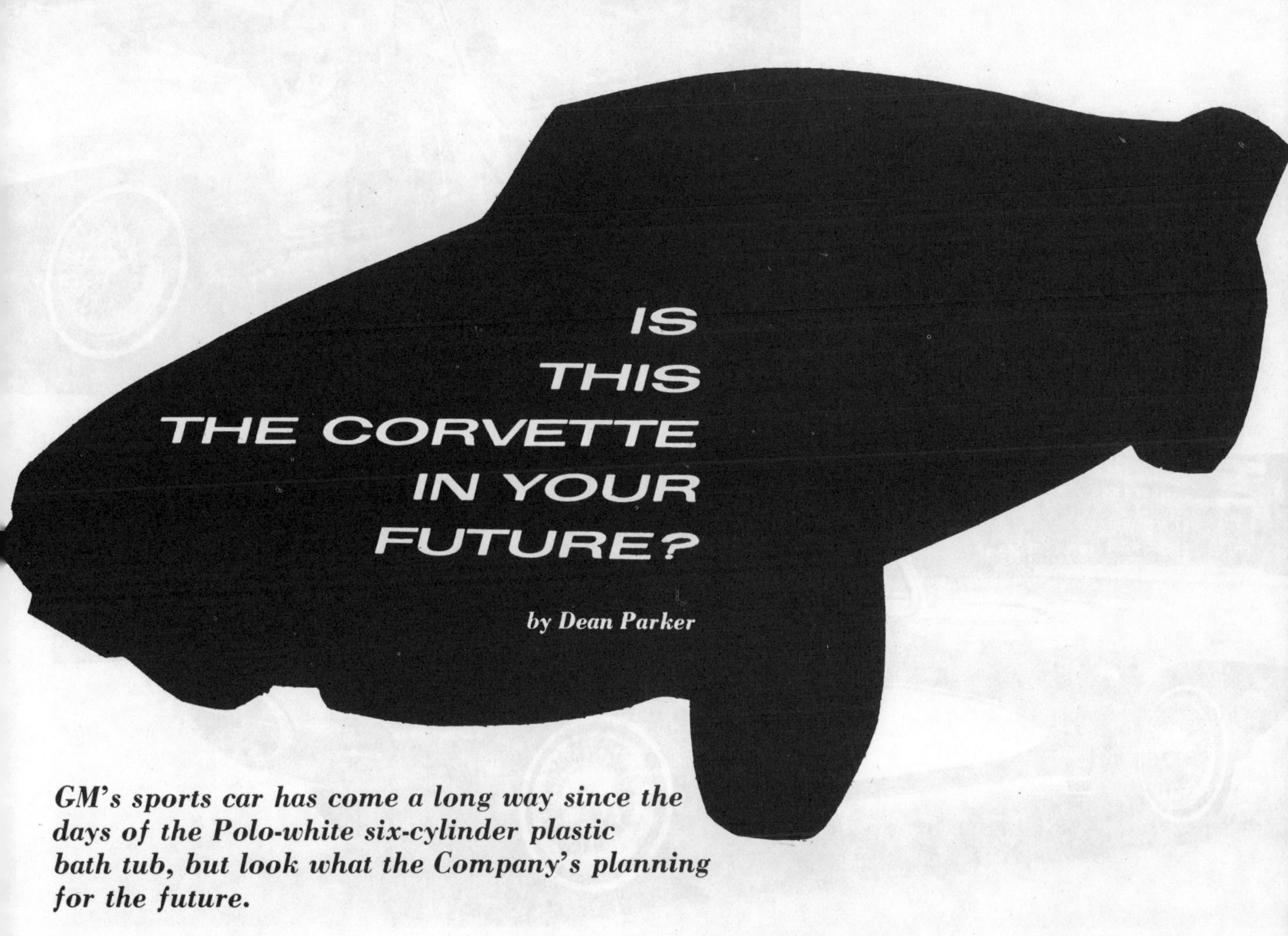

IS THIS THE CORVETTE IN YOUR FUTURE?

by Dean Parker

GM's sports car has come a long way since the days of the Polo-white six-cylinder plastic bath tub, but look what the Company's planning for the future.

If Chevrolet is planning to use steel instead of fiberglass as the body material for the new Corvette, this would involve a redesign of all body dies. There is no reason to believe that the 1960 model will resemble to any great extent the models of the past years. Using as a basis the customized Corvette of Mr. Mitchell, and especially the rear deck and fender treatment, Don Typond has carried this through in the design shown on the following pages. The "wind-split" down the center and the knife edge line around the middle are continuations of the design of the rear end. The Corvette family resemblance is maintained by the use of cut-outs on the sides of the body, and by the familiar shape of the windshield which has been raked back at a greater angle. The flattened hood and front fender line has been suggested by this year's Chevrolet.

THAT CHEVROLET is going to bring out a new Corvette next winter is common knowledge. They've done it every year since the model was introduced. But this year the rumors have gotten wilder than a high-lift camshaft. Aluminum engines, five-speed transmissions, independent rear ends, space frames, aluminum bodies, 400 + cu. in. engines . . . if you believe everything you hear it looks like the sports car to end all sports cars is about to emerge from the Motor City.

Actually the new 1960 Corvette won't be quite this radical. It will be a solid, evolutionary improvement on what is basically a five-year-old design. It will be one of the world's top sports cars, from standpoints of performance, handling and style.

Here is some "well-informed speculation" on what's coming up, gathered from many reasonably reliable sources. Let's begin by establishing that Chevrolet has been thinking seriously about a new Corvette for three years. The current body and chassis has been basically unchanged since the car was introduced in early 1953. The bulk of the tooling investment was paid off long ago. Furthermore, this original Corvette never achieved its intended sales goal – the big market for a "personal" two-seat sports-type car. Ford came in with their better-styled and roomier Thunderbird and all but ran the Corvette right out of the market. Chevrolet had little choice but to concentrate on performance and push the Corvette as an out-and-out sports car. (This was no hardship to project head Zora Arkus-Duntov!) But this is the why behind options like fuel injection, 4-speed transmissions, metallic brake linings, etc.

Under other circumstances this failure to achieve a planned sales goal could've killed the Corvette before it ever got rolling. Fortunately Chevrolet general manager, Ed Cole, was patient. Today, even though the Corvette has never been a big seller, its value to G.M. in terms of free publicity and prestige has been great enough to warrant tooling for an entirely new version. And what's important to you, the performance theme has now become firm company policy. Chevrolet is not going to compete with Thunderbird with their new Corvette. Performance and handling will continue to be the design watchwords. The engineers will consider luxury and big-car roominess—but you can bet they won't sacrifice performance to get it. They'll be more apt to spend a few extra bucks to save weight, and include technical features that would be a waste of money on the mass market.

Two views of the special Corvette built for William L. Mitchell, GM vice president in charge of Styling Staff.

Duntov's crew are on a never ending five-year plan to build a suitable car for U. S. enthusiasts. But because it represents a small percentage of Chevrolet production the Corvette gets only a fragment of each year's re-tooling budget. Many changes will come, but not all at once.

It looks now like the new body will be steel, constructed with new low-cost plastic dies. (These dies are suitable for production rates up to about 25,000 a year, and are much less expensive than steel dies.) The fiberglass body has not proved to be as economical in limited production as originally hoped. The hand layup and vacuum molding process is a nuisance. The body weight of only 340 lbs. is attractive; but Chev engineers tell us the extra beef built into the frame to compensate for lack of strength in the fiberglass structure takes away all the weight advantage. They say they could build a short-wheelbase body for the Corvette out of steel that would carry some of the car structural loads; then a relatively light frame could be used—and the whole car would weigh less than the current fiberglass job, maybe by as much as 150 lbs. if they designed carefully! The bug earlier was the die problem. New plastic dies now make it practical to tool for steel bodies in production rates less than 10,000 a year. This looks like the pattern now.

Stylewise the new Corvette will bear little resemblance to the current model. Wheelbase and overall dimensions will be comparable, but the car will be given a longer, sleeker look by extending the hood forward and down into a Ferrari-like "snorkel" grille opening. Then the front and sides of the body will be shaped to act as huge air ducts for front brake cooling. The fender sections will be cut away at the front to expose the brakes, and the body sides will be swept inward behind the wheels to provide an exit vent for the air flow. The whole deal will be a lot like some late Testa Rossa Ferraris. Actually, some of these rumored styling features coincide suspiciously with a "customized" Corvette that was built last summer by G.M. Styling for their new chief, Bill Mitchell. The car has been on display in the lobby of the Styling building at the Tech Center in Detroit. This is the car, photos of which appear on these pages. It hasn't yet been run on the streets. Draw your own conclusions.

Chassis plans are far from finalized. We know that Duntov was well pleased with the handling of the experimental SS chassis at Sebring in '57, and the word is that he would like to incorporate as many of these engineering features as possible on the new Corvette, within the required cost framework. You can discount space frames, unit or monocoque body construction right off . . . much too expensive. The frame will likely be a channel X, but much lighter than the current design. Front suspension is said to be of the conventional coil spring-wishbone layout, with tubular shock inside the coil, and incorporating ball joints. Some regular Chevrolet passenger car parts could be used here.

The big story would appear to be in the rear end. Persistent rumors mention a DeDion setup on coil springs, with the transmission mounted in unit with the differental. You'll recall the SS used a DeDion rear end, but with the gearbox on the engine. No details have leaked on the wheel control linkage; but the SS setup, with two sets of trailing

Don Typond's impression of how the new Corvette may look.

links for both fore-and-aft and lateral control, looks economical and practical on paper. We hear the rear brakes will probably not be mounted inboard. This would cost some unsprung weight, but might be considerably cheaper to produce. The idea of mounting the transmission at the rear, of course, is to get a more even front-rear weight distribution for traction . . . and, don't forget, the "polar moment" of the car is increased by putting this mass farther from the center of gravity. By thus increasing the "flywheel effect" when turning about the CG, we get somewhat more directional stability at speed and when cornering hard (up to a point). The rear-mounted transmission is the coming thing on all types of cars. Incidentally, we hear no rumor of any new transmission for the Corvette; the current 4-speeder will likely be standard.

Brakes? – no need for improvement here.

And, last but not least, the engine. The next Corvette will not have an all-new engine, as rumored—nor will it use the big 348-cu.in. V8 that's doing such a good job in the passenger cars (see Borgeson's road test in the January '59 SCI). The current 283-cu.in. engine weighs 90 lbs. less than the 348—and that's more than enough reason for the Chev engineers to stick with it for the Corvette. They're serious about this weight problem. In fact, they're doing a lot of experimenting with aluminum to trim the already-low weight of the 283. Remember the experimental aluminum heads for the SS? These were not used at Sebring; but they say the bugs are out of this design now, and there's a good chance they'll be standard on the '60 Corvette. (Remember that, with minor modifications, aluminum can be cast in conventional iron sand-casting equipment.) The water pump casting is also aluminum. The intake manifold will likely be aluminum. They're experimenting with all-aluminum blocks —using an aluminum-silicon alloy that needs no bore liners or plating—but this is not likely for '60. Anyway, with just aluminum heads, manifold and water pump we'd save about 75 lbs.—so the whole engine shouldn't run too much over 450 lbs. Is that worth thinking about??

The other big news item on the '60 Corvette engine is that it's going to be bigger. Chevrolet engineers have cherished their three-inch stroke with a passion—figured it was one big reason why the little 265 and 283 would turn such ungodly engine speeds and get so much horsepower per cubic inch. Could be. But certainly they've pushed the power development of this basic engine so far now (with the special cam, light valves, f.i., etc.) that piston displacement is about the only well left to tap. It is known that Chev engineers are experimenting with bores up to 4 inches and strokes up to 4 inches on the 283 block—which would give 402 cu.in. if combined! (The lower cylinder walls are relieved in the casting stage for rod clearance.) The standard '60 displacement won't be anywhere near this big, but we hear a figure of 332 cu.in frequently mentioned. Of course we can expect the usable rpm to come down when we lengthen the stroke—and this, in turn, will tend to bring down the hp per cu.in. But certainly a rating of, say, 325 hp at 5600 rpm wouldn't be too much to expect from a 332-cu.in. '60 Corvette engine with special cam and f.i. We'll see.

Left: The Stingray and the ex-Sebring Lister Jaguar on the grid during practice.
Below: Dick Thompson brings the car out of the paddock for the start of the feature event at Marlboro.

by Duane Unkefer

THE STINGRAY: THE RETURN OF THE CORVETTE SS

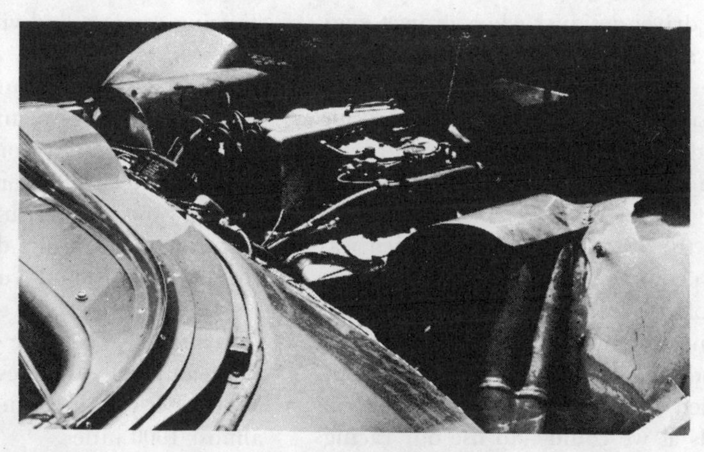

The cockpit (left above) with full-width windshield. When it was found necessary to add a cooling scoop to the rear deck, a section of the screen was cut off to allow air flow (lower right). Upper right photo shows snake-like exhaust headers kicked up to clear space frame. Everybody wanted a close look under the hood but not many questions were answered.

Since its ill-fated debut and subsequent retirement from the racing scene, the SS Corvette, after exhibiting great potential, just seemed to vanish into thin air. A few stories popped up from time to time, including one about the G.M. big brass having ordered its destruction, followed by a cloak-and-dagger routine in which the car was hidden away like the Holy Grail in a successful attempt at eluding the executive Crusaders.

This story may well have been true since on April 18 Maryland's Marlboro Raceway was the scene of the surprise debut of G.M. Styling boss Bill Mitchell's sleek flame-red "Stingray". Right alongside, and serving as Course Marshall's car, was his customized Corvette (SCI March '59) but the Stingray held the spotlight, drawing lots of questions and very few answers.

The flawless and rather way-out fiberglass bodywork of the Stingray bears a remarkable resemblance to its namesake. The chassis is that of the third, previously incomplete Corvette SS, built back in 1957. A full tech report of the SS appeared in the August, '57 issue of SCI.

The frame is built up of tubular members about an inch in diameter, with certain parts square sectioned wherever it was found more advantageous for purposes of mounting other components. The result is a space frame quite similar to that of the 300 SL even to the pyramided tubes at the cowl and truss structure at the sides of the cockpit.

The engine looks like the same one that was used in the Sebring car, but then all Corvette engines look pretty much alike from the outside. It is rumored that all the goodies on the inside are readily available over the counter at your local Chev shop, if you have the desire and knowledge to know what to pick out of the parts bin. Stock, man, is stock. And there are various degrees of stock. The gearbox is probably the same aluminum four-speed box that was used two years ago as there is no reason to suspect a change; the Sebring car's problems did not lie in the gears.

CONTINUED ON PAGE 100

Driver Dick Thompson (left) and Zora Arkus-Duntov (right) give the author information about the car. Duntov made the understatement of the year... "It's just a cobbled-up special."

► No one knows Corvettes like Zora Arkus-Duntov, the highly practical driver-designer who's project engineer for Chevrolet's sports car. His views on the development of the latest Corvettes are of great import: "Originally, our plan was to develop the car along separate touring and racing lines, as Jaguar did with the XK series on one hand and the C-Type and D-Type on the other. With this in mind we first introduced racing options, then the SR2, finally the SS, which was intended to be our 'prototype' competition car. When this project was cut off, we realized we had to approach the Corvette in some other way. Since we could no longer build two kinds of Corvettes with different characteristics, we decided to give the Corvette buyer as much of *both* worlds as we could – to use our racing experience to combine in one automobile the comfort of a tourer and the ability of a racer. A big order, yes, but an interesting and worthwhile one. The 1960 Corvette was the first to reflect this thinking; the 1961 car is very similar."

Before talking to Mr. Duntov about the 1961 Corvette, SCI had formed this strong general impression of it: one of the most remarkable marriages of touring comfort and violent performance we have ever enjoyed, especially at the price. That our impression matched Chevy's intention so exactly is a tribute to the job done by Duntov and his crew. We had ample time to sample the latest Corvette and get to know its virtues and vices, yet could complete the rigorous R.R.R. routine in time to publish this report scant weeks after its introduction. Unique dispensation by Chevrolet made this possible. In mid-August, well before official release, we picked up production Corvette number three in Detroit, drove it to New York for testing at Lime Rock Park and our other test areas, and returned it to the Motor City. Altogether SCI drove this automobile almost 1900 miles.

PLENTY OF POWER

Since this red and cream machine was equipped with the hottest engine on the long list of Corvette options, it's no surprise that we were stunned – as in earlier Corvette tests – by its terrific performance. It was equipped with the Duntov cam and Rochester fuel injection, the latter always being accompanied by

Road Research Report: CHEVROLET Corvette

special cylinder heads. Therein lies a tale. As is well-known, last year Chevrolet introduced aluminum cylinder heads for the fuel-injected engines. In addition to the change in material, there were important design alterations which accounted for the gain of 25 bhp over the previous 290 bhp rating. For better breathing the

intake valve head size was increased from 1.72 inches to 1.94 inches, which was helpful but which couldn't boost power unless matching improvements were made in the intake port. It couldn't be enlarged all the way through because the pushrods pass close to the ports near their outer ends, so the port was necked down to a venturi shape adjacent to the pushrod holes, then allowed to expand smoothly on its way to the combustion chamber. It works very well.

The alloy used for these 1960 heads contained a high percentage of silicon, an amount not specified by GM but certainly in the 16 to 20 percent bracket – sufficient, anyway, to make inserted valve seats unnecessary. The casting method used was the same low-pressure system now used for the Corvair heads and other parts, in which the molten alloy flows up into the mold from below, under moderate pressure. The casting is an intricate one, and the high-silicon alloys tend to be difficult to handle, with the unfortunate result in Chevy's case that there were frequent faults in the castings and consequently a high rate of rejection. It did happen that an occasional head managed to conceal a flaw through all the inspections and smuggle it aboard a Corvette; this was one of the two prime causes of the field failures that gave these heads a poor reputation. The other cause was overheating. If for some reason a Corvette's cooling system failed or lost water, the aluminum heads, with their lower melting point, were much more likely to be damaged than were iron heads.

If you had a set of sound aluminum heads, then, and kept your cooling system in good order, you had absolutely nothing to worry about. Many Corvette owners racing with aluminum heads today will bear this out. But that high rejection rate, plus the frequency of field failure, led Chevy to decide to stop supplying the heads in 1961, and to replace them with cast iron heads that incorporate the same refinements to the intake tracts. These iron heads were given their baptism on the Cunningham Corvettes at Le Mans last June, and are now supplied on the 1961 fuel injection engines.

INJECTION REFINEMENTS

To get those added horses at the top end, this engine obviously has to breathe more air, which could be supplied by increasing the area of the intake venturi. Doing this, however, would weaken the metering signal at low and medium speeds, much the way a large carb venturi brings on low-speed metering problems. Duntov says that the extra air was needed for short periods only at very high engine speeds so Chevy built in a reserve supply by increasing the height of the "dog house" of the fuel injection unit. The fins were trimmed off the top, which was raised almost an inch to increase the volume of the plenum chamber inside. (The cross-section on page 46 shows the 1957 engine with the finned manifold, drawings of later engines being unavailable.) The ram pipes keep their original 12-inch length in the course of this change, which was made in 1960. Another important change in the engine department is the use on *all* 1961 Corvettes of the Harrison aluminum radiator fitted only with the Duntov-cam engines last year.

Once you know how, starting the fuel-injected Chev engine is a snap either cold or warm. In the former case you leave your foot off the gas; in the latter you press the pedal to the floor. A fast-idle setting is in effect when the engine's cold, and, like any good injected engine, it runs smoothly and regularly right from the start. As smoothly and regularly as it ever will, in this case, for the warmed-up 850-rpm idle is definitely rough. With solid tappets and 66 degrees of overlap this is hardly surprising, and Duntov points out that the post-1960 increase in intake valve size had the effect of aggravating the slight roughness that existed earlier. It also makes the unit a little easier to stall if you don't apply enough gas. Throttle response is excellent – instant, proportional to pedal position, and cutting off power effectively on the over-run.

Above we're talking about response in terms of control. Response in terms of a kick in the back is sudden and convincing. The broad range of power offered by this remarkable V8 is extremely impressive; credit must be given the big valving at the top end and the injection over the rest of the range. No matter what gear you're in or what speed you're going, you can step down hard and get the same surging lunge forward. As the Engine Flexibility graph shows, the powerplant pulls smoothly and strongly from idle right on out to the red line, which is marked on the tach at 6500. We used 6500 for our acceleration runs, though Duntov feels you get equally good results by shifting at 6000. So well are these special-cam engines assembled and balanced that you can, if necessary, look at 7000 or even 7500, something we didn't try and Chevrolet didn't encourage.

SOUND AND FURY

Appropriately for a sports car, this engine is exciting to listen to. Along with the new back end shape, the twin tail pipes have been redirected so they angle outward just behind the rear wheels. They rumble with a truly musical motorboat tone and beat a tattoo on the sides of the cars you (frequently) pass. As the revs rise, the tone ranges up through several octaves to a musical moan that emanates more from the engine than from the exhaust. In short: a hard, solid, machine-like sound that inspires confidence. At or near idle speed, especially with the top up, a vibration period materialized on our test Corvette that set up a "thrumming" wihin the cockpit. The new, more forward location of the exhaust outlets also seemed to make the exhaust more audible in the cockpit than on earlier versions, something that makes you think the engine is revving faster than it should at cruising speeds. The sound can become as tiring on a long trip as it is fun on a short one.

By any standard at all, the Corvette we tested is a sensational performer. Right up to 60 mph, in fact, its curve traces an arc much like that of the 250/GT Ferrari (The Corvette took 6.7 seconds against the Ferrari's 6.6). Unlike the Ferrari, the Corvette goes all the way to 60 in first gear, then after the shift to second the curves begin to diverge as the Ferrari's lighter weight and superior aerodynamics make their effect felt. Yet the Ferrari had a vibrant, restless air about it that was constantly urging you to put your right foot down and *go.* Not so the Corvette, which is just as happy pottering along byways at more sedate velocities. On this car, there were four speeds available but it never seemed to matter which one we were in. Our trips back and forth to Detroit could have been made all in first gear or all in fourth gear, for example, still keeping up with the traffic. Just for amusement, we took some rough acceleration times to sixty in each of the gears, using one gear only each time. In second it took about 8 seconds, in third about 10, and in fourth gear only a little less than 15!

BOX AND BRAKES

It is certainly not the intent of Chevrolet that you should not shift this transmission, however. It's one of the quickest, slickest shifts available today on any automobile. Its short, plastic-knobbed lever is spring-loaded to the right side of its very narrow gate, and the

Text continued on page **54** – *data overleaf*

Dechromed headlights and a thin-line grille give the front a new look too. Push on front edge opens lid, below. We had to try it every 160 miles or so.

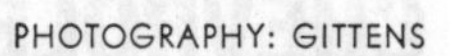

PHOTOGRAPHY: GITTENS

Cockpit continues emphasis on high style. Subtle changes over the years have brought more leg room, improvements in driving position. Seats look more buckety than they are, offering little side support.

Corvette requires meaningful coordination of steering and throttle on corners. Limited-slip rear end lets you use power to hold it on line.

Upswept tail resembles XP-700, ties design to Corvair styling. Exhaust pipes end behind wheels, shouldn't get bent on steep driveways or curbs.

Road Research Report:
CHEVROLET Corvette

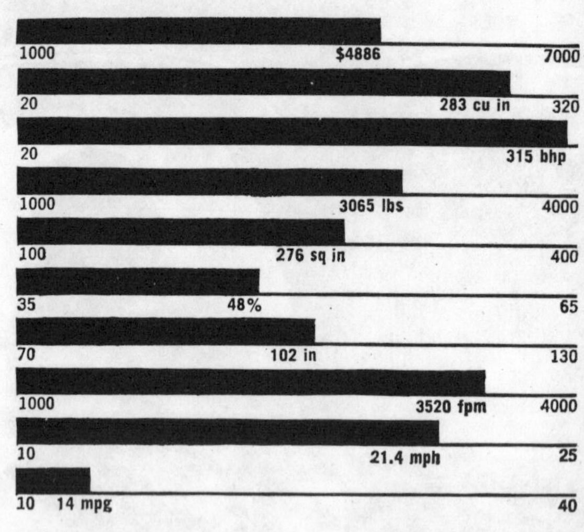

Item	Scale min	Value	Scale max
Price as tested	1000	$4886	7000
Displacement	20	283 cu in	320
Power (SAE)	20	315 bhp	
Curb Weight	1000	3065 lbs	4000
Swept Braking Area	100	276 sq in	400
Weight on Driving Wheels	35	48%	65
Wheelbase	70	102 in	130
Piston Speed, "corrected"	1000	3520 fpm	4000
Speed @ 1000 rpm in Top Gear	10	21.4 mph	25
Mileage	10	14 mpg	40

Manufacturer: Chevrolet Motor Division General Motors Building Detroit 2, Michigan

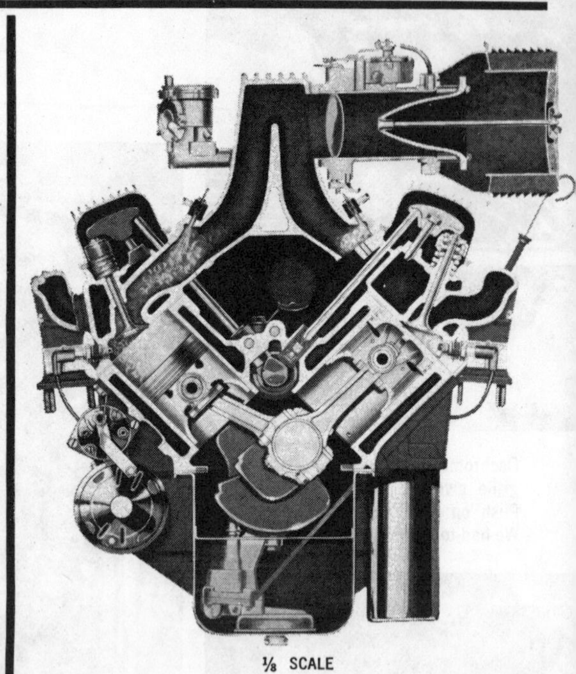

1/8 SCALE

ENGINE:

Displacement ....283 cu in, 4640 cc
Dimensions ....Eight cyl, 3.875 x 3.00 in
Compression Ratio ....11.0 to one
Power (SAE) ....315 bhp @ 6200 rpm
Torque ....295 lb-ft @ 4900 rpm
Usable rpm Range ....850-6500 rpm
Piston Speed $\div \sqrt{s/b}$ @ rated power ....3520 ft/min
Fuel recommended ....Premium
Mileage ....10-18 mpg
Range ....165-290 miles

CHASSIS:

Wheelbase ....102 in
Tread, F,R ....57, 59 in
Length ....178 in
Suspension: F, ind., coil, wishbones, anti-roll bar; R, rigid axle, leaves, radius rods, anti-roll bar.
Turns to Full Lock 1.9
Tire Size ....6.70 x 15
Swept Braking Area ....276 sq in
Curb Weight (full tank) ....3065 lbs
Percentage on Driving Wheels 48%
Test Weight ....3390 lbs

DRIVE TRAIN:

Gear	Synchro?	Ratio	Step	Overall	Mph per 1000 rpm
Rev	No	2.26		8.36	9.5
1st	Yes	2.20		8.14	9.7
			33%		
2nd	Yes	1.66		6.15	12.9
			27%		
3rd	Yes	1.31		4.84	16.3
			31%		
4th	Yes	1.00		3.70	21.4

Final Drive Ratios: 3.70 to one standard; 3.36, 3.55, 4.11 and 4.56 optional.

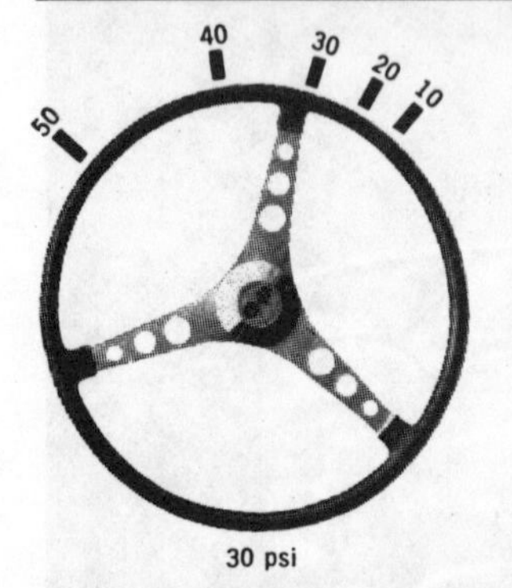

Steering Behavior

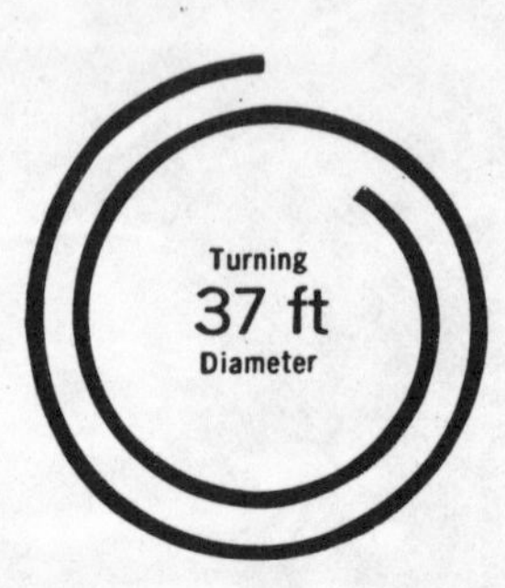

Turns to Full Lock

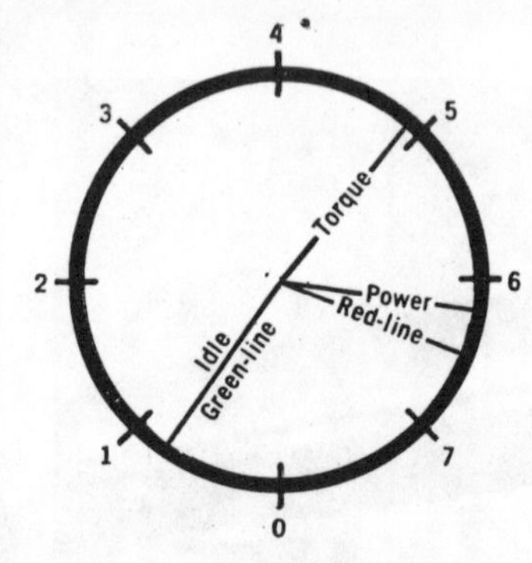

Engine Flexibility

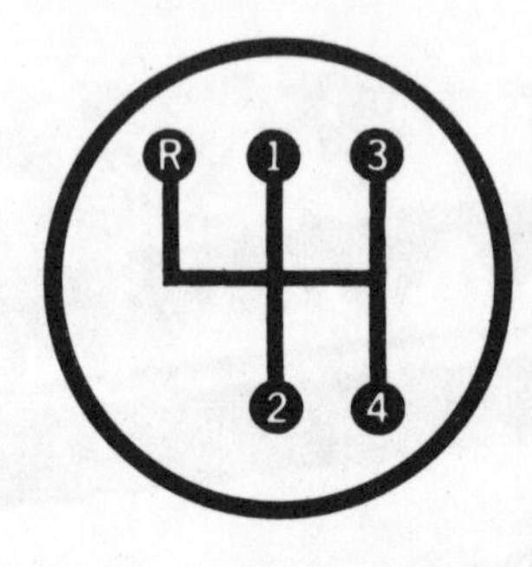

Shift Pattern

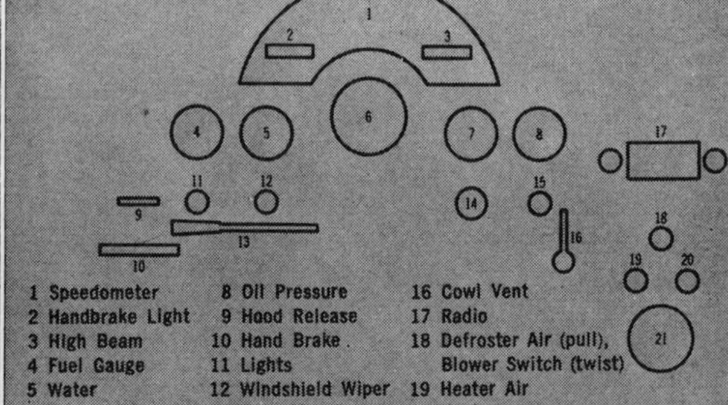

1 Speedometer
2 Handbrake Light
3 High Beam
4 Fuel Gauge
5 Water Temperature
6 Tachometer
7 Ammeter
8 Oil Pressure
9 Hood Release
10 Hand Brake
11 Lights
12 Windshield Wiper
13 Turn Signal
14 Ignition
15 Lighter
16 Cowl Vent
17 Radio
18 Defroster Air (pull), Blower Switch (twist)
19 Heater Air
20 Heater Temperature
21 Clock

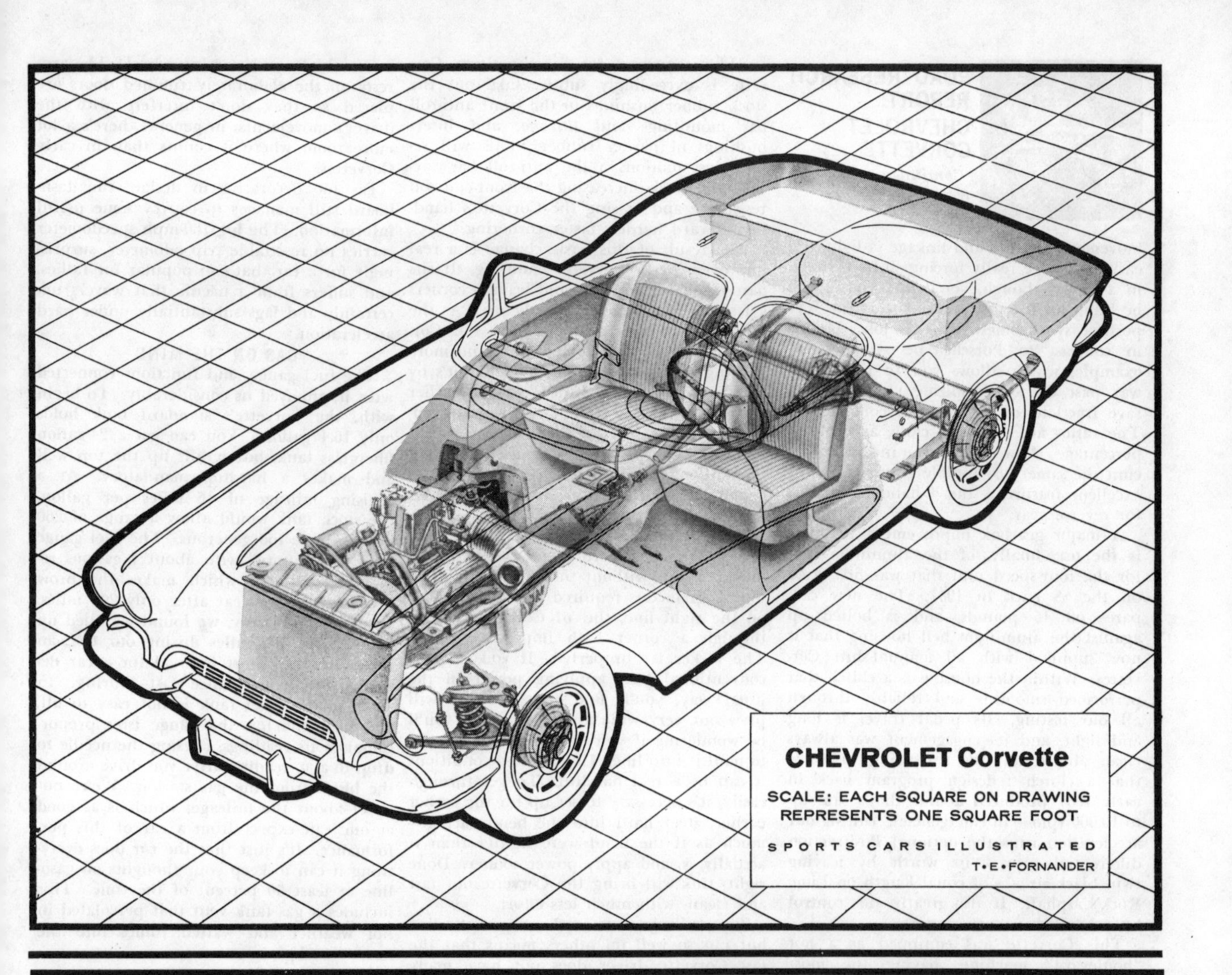
CHEVROLET Corvette
SCALE: EACH SQUARE ON DRAWING REPRESENTS ONE SQUARE FOOT
SPORTS CARS ILLUSTRATED
T·E·FORNANDER

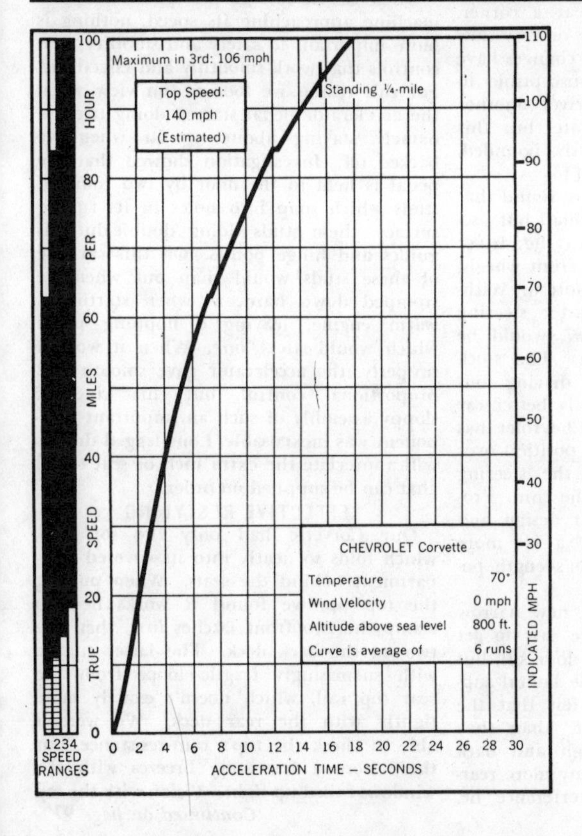
Maximum in 3rd: 106 mph
Top Speed: 140 mph (Estimated)
Standing ¼-mile
TRUE SPEED MILES PER HOUR
INDICATED MPH
ACCELERATION TIME – SECONDS
1 2 3 4
SPEED RANGES
0 2 4 6 8 10 12 14 16 18 20 22 24 26 28 30
20 40 60 80 100
30 40 50 60 70 80 90 100 110
CHEVROLET Corvette
Temperature 70°
Wind Velocity 0 mph
Altitude above sea level 800 ft.
Curve is average of 6 runs

ROAD RESEARCH REPORT: CHEVROLET CORVETTE

Continued from page **50**

movement of lever and linkage is light and easy. If you're really forcing shifts through in a hurry, though, you'll find that you'll be hung up for just an instant as the very positive synchromesh does its job. This is in contrast to Porsche-type synchro, for example, which allows you to push your way past the synchronizing mechanism to save fractions of a second in competition. The ratios are high and close, as the step percentages show, and all the indirect gears emit the same, just-audible gear whine. An excellent feature is the T-handled lockout for reverse gear.

A major gearbox improvement for 1961 is the use, finally, of the aluminum case for the four-speed unit that was first tried on the SS back in 1957. The new case pares off 15 pounds, and is bolted up against the aluminum bell housing that is now supplied with all manual-shift Corvettes. Within the housing is a clutch that performed smoothly and reliably through all our testing. Its pedal travel is long and light, and its engagement was always clean and solid. It's reassuring to know that a clutch redesign program back in early 1957 produced a unit that's safe up to 10,000 rpm — in emergencies. During our acceleration runs the optional Positraction differential proved its worth by leaving twin black streaks of equal length on Lime Rock's asphalt. It also greatly aids control by power when cornering.

This Corvette was equipped as a hot "boulevard" machine, having the most potent engine but with standard brakes and tubeless whitewall tires. Bringing the car to a halt after a few of those scalding 0-100 runs quickly induced fade that was marked enough to cause alarm. Fortunately, the brakes recovered very quickly, with no grabbing, pulling or other signs of damage. But they are definitely not nearly up to the car's performance. To get suitable brakes, you'd select Regular Production Option 686, which combines sintered iron linings with the standard drums. RPO 687 goes all the way with a finned, fan-equipped drum and scoops in the backing plates, still using the sintered iron linings. The grabby, unpredictable ceramic-metallic linings are now a thing of the past, having served their purpose as a stopgap until Moraine could design sufficient life and consistency into the sintered iron type.

IMPROVED HANDLING

Finally, at the end of its development life, the original Corvette has been endowed with handling that allows its ample power to be used for control, in both street and race track trim. How? By making street and race chassis trim identical! This policy began in the 1960 model year, when the optional stiff suspension kits were discontinued, and the standard suspension was stiffened in roll by removing the original kink in the 13/16-inch front anti-roll bar and by fitting a 5/8-inch anti-roll bar at the rear. This trim is unchanged for '61. The only suspension change that Duntov would recommend for a competition Corvette is exceedingly subtle: take out the stock rubber bushings in the front anti-roll bar mountings and linkage, and insert bushings of harder rubber. This will reduce lost motion in the anti-roll bar's action, effectively increasing the front-end roll resistance and biasing the Corvette's handling toward harder, faster cornering.

The result of this basic change is a real improvement in Corvette handling. It sits flatter, making entering and leaving corners a less tricky maneuver than before. As the Steering Behavior graph shows, the Corvette understeers, requiring more and more steering lock as speed increases, but it's by no means as gross an understeerer as earlier versions, which (in street trim) would simply plow right off the road as power was applied. Now power can be used to bring the tail end out when cornering fast, reflecting a much better balance between front and rear suspensions.

When either end of the Corvette does swing out in a corner, it swings pretty quickly and without much warning, so some vigilance is required. Properly set up on the right line, the '61 Corvette can get through a corner with impressive speed. The kicker is "properly". If you select a conventional line, applying power in the usual way, you'll find that the car will plow out very wide at the exit and you'll be wondering if you have strength enough to haul it into line. Under these conditions it can be a real handful. To get through easily, it's necessary to set up for the corner earlier, steer hard into the bend earlier — much as if the bend were tighter than it actually is, and apply power sooner. Done right, this will bring the Corvette out fast and clean, with much less effort.

This willingness to work well on one line but not so well on others means that the fast Corvette driver does not have many choices of position, in or near a corner. He's either on the proper line or very slow or in trouble. The suspension changes have made the car much more susceptible to power control, giving the driver another dimension in which to operate, but this new realm is still very strictly bounded.

BEHIND THE WHEEL

During our handling tests, we found that the steering now transmits a small but useful amount of information to the driver regarding the adhesion at the front wheels, and has a minimum of lost motion. Without a doubt, the slightly faster steering ratio, supplied with RPO 687, would be just right for fast cornering. The stock ratio is slow for sporting driving but fine for touring. It's a basically better car to drive now, also, because Chevrolet has steadily improved the driving position over the years. No longer must the steering wheel be placed right under the chin. Progressive improvements in seat design and travel have given the driver a lot more room, if not a continental arm's-length position.

Trimly upholstered with a new narrow ribbing, the Corvette seats are easy to get into and out of, over a wide door sill, but offer (as a corollary) very little lateral support to the occupants. We felt that the seats were placed more "flat" than they might be for maximum thigh and back support, but we realize that any more rearward angle would cause interference between thighs and steering wheel. The arm rests on the elaborately-trimmed doors are placed so they don't interfere with the driver's movements; in general there's a lot more room where it counts than in early Corvettes.

Far too decorative in design, the dashboard still manages to convey some useful information. The big 160-mph speedometer carries no resettable trip odometer, surprisingly for a car that's so popular for rallies, and suffers from a needle that wavers uncertainly and lags substantially under hard acceleration.

GAS ON THE MIND

The fuel gauge, and functions connected with it, annoyed us considerably. To begin with, the Corvette's standard tank holds only 16.4 gallons. You can get a 24-gallon fiberglass tank, but it fills up the top well and makes a hardtop mandatory. At a cruising mileage of 16 miles per gallon, the stock tank would allow a range of 260 miles — none too generous. The fuel gauge indicates "empty" with about 4 gallons remaining, though, which makes the brow break out with sweat after only 200 miles. Speaking practically, we found we filled up about every 160 miles during our test, an inconveniently short interval for a car designed for long-distance fast touring. In addition, the fuel tank is not easy to fill, and near the top the gauge isn't proportional in its readings, causing the needle to drop to a ¾ reading after you drive around the block from the gas station. We're not upset about the mileage, which is as good as one can expect from a car of this performance. It's just that the car does everything it can to keep your thoughts on gasoline at least 50 percent of the time. This includes a gas tank vent that percolated in hot weather and wafted fumes into the nearby cockpit.

On a car as fast as this one, or even a machine approaching its speed, nothing is more important to safety and stability than controls that work smoothly and effectively. For this reason we took a dim view when the accelerator pedal started doing tricks — namely staying about ¾ on when we backed off. Investigation showed that the pedal is held to the floor by two rounded studs which snap into holes in its rubber surface, these studs doing double-duty as guides and hinge points. On this car one of these studs would snap out when we tromped down hard, as when starting a warm engine, leaving a flopping pedal which would stick "on". When it worked properly, the accelerator gave smooth and proportional control, but this kind of sloppy assembly of such an important component was inexcusable. Long-legged drivers will appreciate the extra inch of seat travel that can be supplied on order.

EFFECTIVE RESTYLING

Our Corvette had only the soft top, which folds so neatly into its covered compartment behind the seats. When putting the top up, we found it works best to clamp the two front latches first, then the two on the back deck. The latter engage with surprisingly fragile loops from the rear top rail, which doesn't exactly mate tightly with the rear deck. We weren't able to check the top's rain-resistance, but there were a few stray breezes with the windows rolled up tight. Vision with the top

Continued on page **97**

STING RAY

by Karl Ludvigsen

The fish-like snout of Bill Mitchell's Corvette-SS-based racing car has been nibbling at the exhaust pipes of some highly esoteric iron lately. It shows what can be done with a careful financial outlay plus a lavish amount of engineering development, time, and driving talent.

"It's not a four-alarm advance over all existing equipment, but it is basically a good car. Given more than half a chance and some intensive track testing it could compete with the world's best." That's what SCI said in the August, 1957 issue about the Corvette SS, Chevrolet's answer to the 450S Maserati, which caused such excitment at Sebring in '57. Then it seemed that the de Dion-reared, space-framed SS had been sunk by Chevrolet management, foundering in the wake of AMA's racing ban. It seemed that way for two long years, until April 18, 1959, when a brand-new sports-racing machine of startling appearance was decanted from its trailer at Marlboro, Maryland and driven to a fourth overall by Dr. Dick Thompson. It was, as the July, 1959 SCI reported, "The Return of the Corvette SS."

Since that day the SS, reincarnate as the Sting Ray, has adequately proved our point of August, 1957, thanks largely to the longstanding automotive enthusiasm of Bill Mitchell, GM's vice president in charge of styling. Harley Earl, Bill's predecessor in that job, used to say that a good automobile stylist had to have "gasoline in his veins." That's certainly true of Mitchell, who has been keenly interested in competition cars since he found out about them almost thirty years ago. Finally financially able to field a racing team himself, he looked around for suitable equipment – preferably GM-based, of course. His eye fell on the Corvette SS.

To backtrack a bit, recall that Chevrolet designed the sports-racing SS in a dramatic crash program between September and November, 1956, and had completed the first test chassis at the end of the following January. With a rough fiberglass body this became the "Mule", which went down to Sebring for on-the-spot trials while the actual race car was completed.

Crisp lines of Sting Ray are a rakish blur at Bridgehampton, where Dick Thompson finished second overall behind Walt Hansgen's Birdcage.

After Sebring the Mule was revised and cleaned up in detail to be exactly like the race SS, but the ax fell on the project before the ex-Mule could be assembled and the planned two-car team prepared for Le Mans. It was this car, the Mule that Fangio and Moss drove in Sebring practice, that Bill Mitchell remembered. Bill was able to take the chassis over, on the obvious condition that it be re-bodied and raced as Mitchell's own private project, divorced completely from GM. This is the way it was done, starting with Bill's own design for the new body, in 1958, and construction of the Sting Ray over the '58-'59 Winter.

THE WHO AND HOW

Obviously a lot of the specialized skills of the General Motors Styling Department were utilized in building the superbly-detailed Sting Ray, but all the subsequent racing and maintenance expenses have been entirely Mitchell's. The car is garaged and worked on in a cubicle in the back of the A. J. Berna Company, a woodworking shop in Roseville, Michigan — well east of the GM Tech Center. As of June, 1959, at Elkhart Lake, responsibility for management of the one-car team has been in the hands of Dean Bedford, Jr., a highly canny enthusiast who also supervises the engineering development of the Sting Ray. In 1959 a lot of the mechanical work was done by Eddie Zalucki, but since Nassau of that year R. Ken Eschebach has single-handedly maintained and modified the car. It's a tightly-knit team, closely geared to the needs and preferences of Sting Ray driver Dick Thompson. The number of races that could be entered — hence the pace of development of the car — has been restricted by the limited depth of Bill Mitchell's pocket-book. Nevertheless the Sting Ray's performances in 1959 and 1960 have shown once more that the Corvette SS was at least a match for its contemporaries and would have been no mean competitor today.

When the broad, spatulate Sting Ray body was first created, it was of fiberglass of .125-inch thickness, about the same as on the Corvette. It was reinforced with aluminum, annealed to be soft enough for easy forming; it proved to be too soft to stand up well. After the Road America 500 in 1959, the decision was made to recast the front and rear body sections in lighter fiberglass (.060 inch) before Nassau. This was done by using only three layers of fiberglass silk, reinforced this time with balsa wood! Various hinge and attachment points are of aluminum, heat-treated for strength and bonded in place. This change alone, in just those two sections, saved 75 pounds of weight, and gave the body a flexibility that allows it to "give" when struck instead of cracking and breaking as it did with the heavier thickness.

The big up-folding hood section is reinforced by an aluminum box across the nose and by a tube across the cowl that anchors the fasteners and stiffens the hood. Wherever Dzus fasteners are used in conjunction with fiberglass, which is frequently, the 'glass is given only a slightly thicker section around the Dzus screw head. No heads have ever pulled through. With the SS one major problem was heating of the cockpit by the high-flying exhaust piping. The Sting Ray has a less flamboyant pipe layout, more thorough shielding of the pipes, and a firewall insulated by a half-inch layer of Microquartz, a remarkably lightweight insulation held in place by mesh screening.

STYLING WITH A REASON

Though the body's shape is obviously styled for visual impact (likely to be as pronounced in the front of the '62 Corvette as it is in the back of the '61, we understand), some aerodynamic principles were under test as well. By making the top of the body relatively flat and deeply bellying the bottom, it was hoped that an inverted-airfoil shape could be created that would press the car more firmly against the road. Tunnel testing, both of a model and of the actual car, has shown a drag coefficient of .43 to .45 (not particularly low) and,

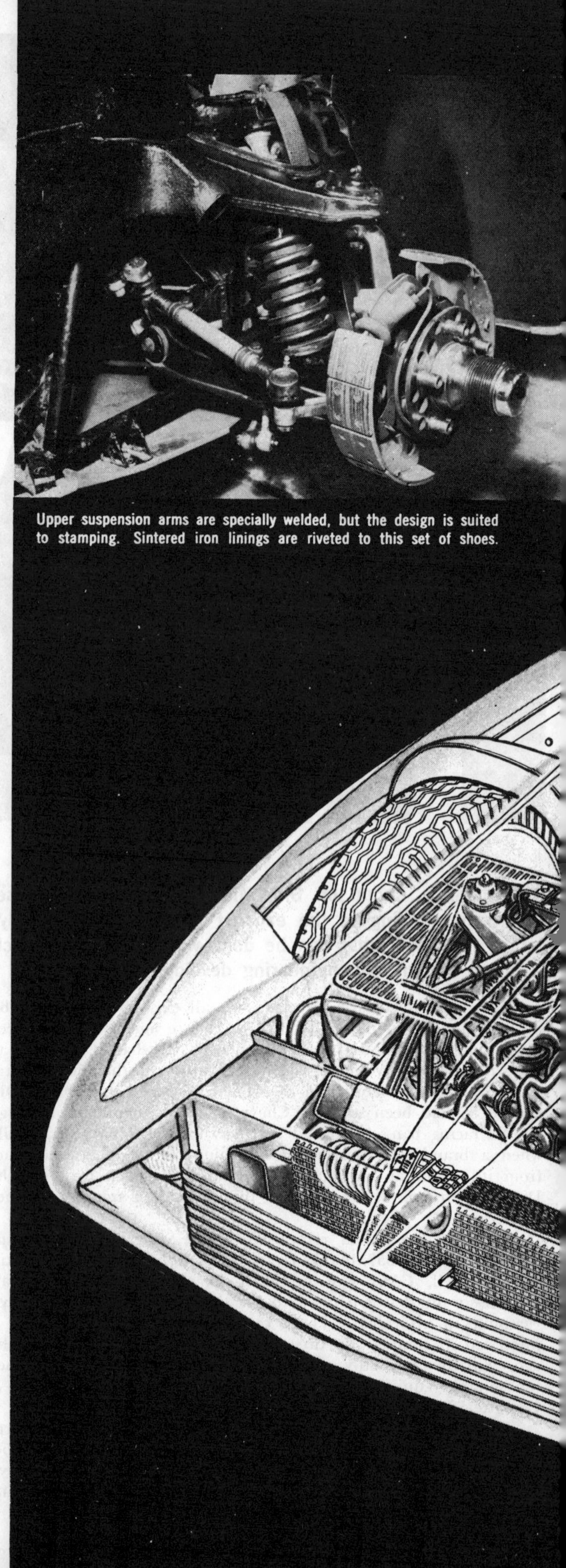

Upper suspension arms are specially welded, but the design is suited to stamping. Sintered iron linings are riveted to this set of shoes.

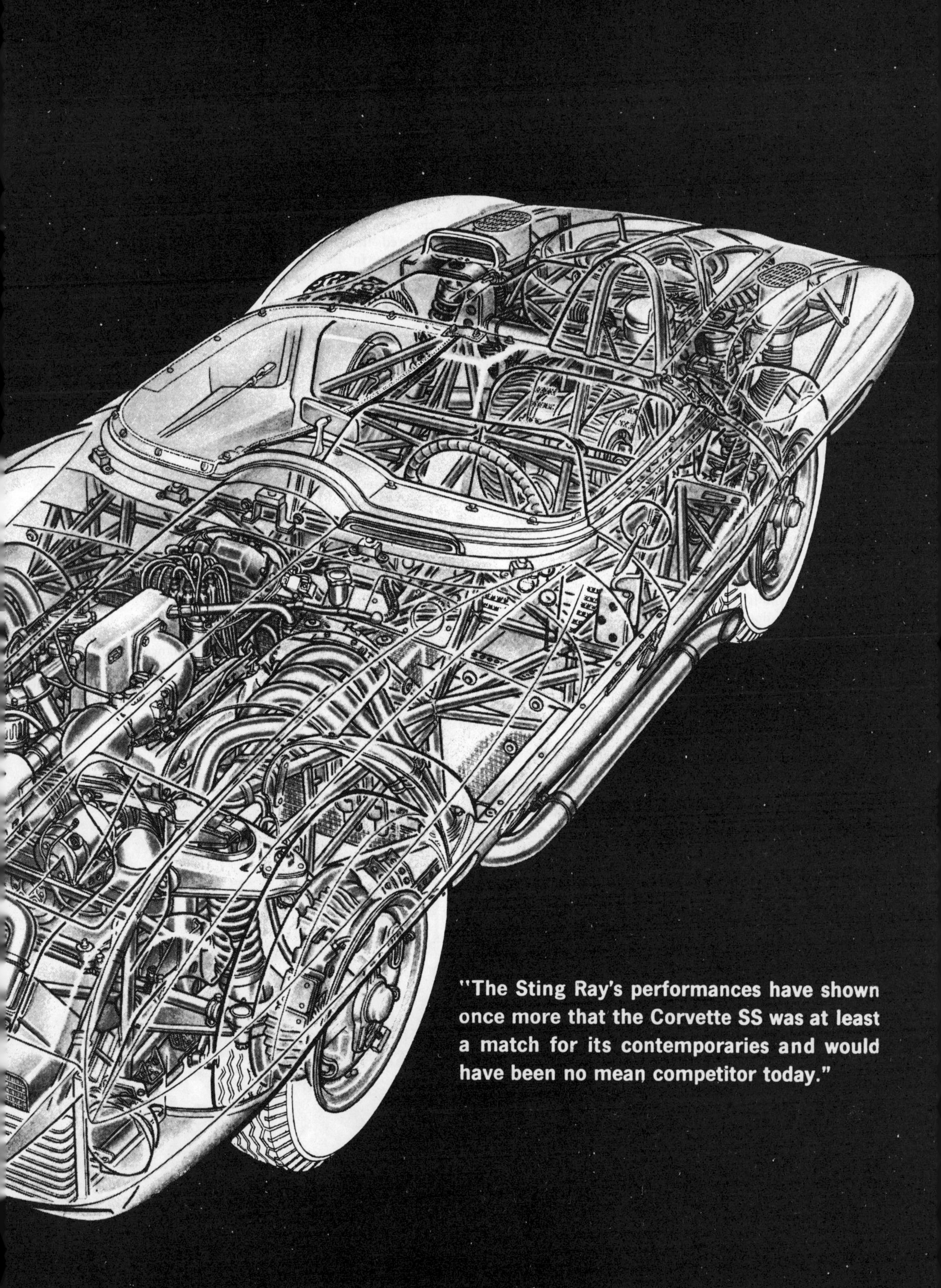

"The Sting Ray's performances have shown once more that the Corvette SS was at least a match for its contemporaries and would have been no mean competitor today."

in 1959 trim, a certain amount of *lift* that had hampered high-speed cornering through '59. By placing shims above the rear coil springs but not the fronts, the body was given more forward rake and the lift markedly reduced.

When he drove the original SS at Sebring, Piero Taruffi said he thought the springs were too soft all around, causing excessive roll and allowing bottoming over bumps. Following through on this, Dean Bedford shimmed up the springs all around in 1959, then changed to ten-percent-stiffer coils in 1960. This was a big improvement, and another ten percent more stiffness may be tried soon. Delco shocks are used all around, as on the SS, and the front anti-roll bar is exchanged for a lighter bar for racing on very tight tracks, to reduce the car's understeer.

Another Sebring bugaboo was the rubber bushings used in the rear suspension radius rods; these are from the Plymouth front suspension and were not inclined to stay in position back in '57. They still tend to break up under the heavy loadings of competition and are now changed as a matter of course every third race. If the car were to be revised basically for coming seasons, Bedford would consider replacing these bushings with ball joints — but he points out this isn't an easy thing to do. There's too much weight in the Sting Ray's nose for it to handle extremely well, but it still gets around a track with remarkable speed. Of course Dick Thompson is entirely at home in the machine now, and it's genuinely exciting — and a little awe-inspiring — to watch him hurl this car through bumpy, hilly bends like those at Bridgehampton or V.I.R. The much-used word "hairy" was never more applicable!

PROBLEMATIC BRAKING

No Sting Ray problem has been more persistent than the brakes. As you'll recall, the SS was equipped with an unique Kelsey-Hayes double-booster system designed to prevent rear brake lockup under high deceleration, using a mercury switch in an involved arrangement of wires and piping described completely in our SS story. One of the '57 problems was too-small line diameters that caused a delay in the system's response; this was rectified then but there still seemed to be a slight lag in the Sting Ray's reaction times, leading to uncontrollable brake lockup — usually at highly critical moments.

At first this lockup was attributed to the combination of the often-erratic cerametallic linings with the Chrysler Center-Plane brake mechanisms that were and are being used, these mechanisms being the two-leading-shoe type which is prone to locking if linings aren't just right. First step, then, was to start using the Moraine sintered-iron linings now fitted to competition Corvettes. Tried first at Lime Rock in July of '59, these linings are used with stock cast iron Corvette drums that are given a fine surface finish and closely matched to the individual linings. At first the sintered linings were riveted in place, then after Bridgehampton in 1960 they were welded to the shoes. During this period the Delco-Moraine Division has experimented with many different sintered iron lining compositions on the Sting Ray, contributing ultimately to the consistency of the Corvette's linings today.

This wasn't the whole answer, though, and the Kelsey-Hayes system was finally thrown out after the Road America 500 in 1959 and replaced by a single Hydrovac booster, modified by Bendix to allow changes in pedal pressure to be made. The brakes are still not very impressive, which leads to a logical question: why not discs? This falls in the category of the ball joints for the rear suspension; it would be nice but it's a very big job, involving extensive changes in the hub layout to keep the discs from causing an excessively high kingpin offset. The Sting Ray's weight is naturally a factor here. With the old body and original brake system, the car weighed 2154 pounds dry, leading to an estimate of a present dry weight of 2000 pounds. Wet, with the 31-gallon normal fuel tank full, it should weigh about 2360 pounds. It's no Birdcage, but neither is it disgracefully heavy.

MODEST POWERPLANT

The Sting Ray's excellent performance in its class (C Modified) is surprising in view of its "cooking" engine, as Stirling Moss described the SS unit at Sebring. In fact the Sting Ray crew doesn't even claim as much power for it as the latest hot production Corvettes are boasting! They say

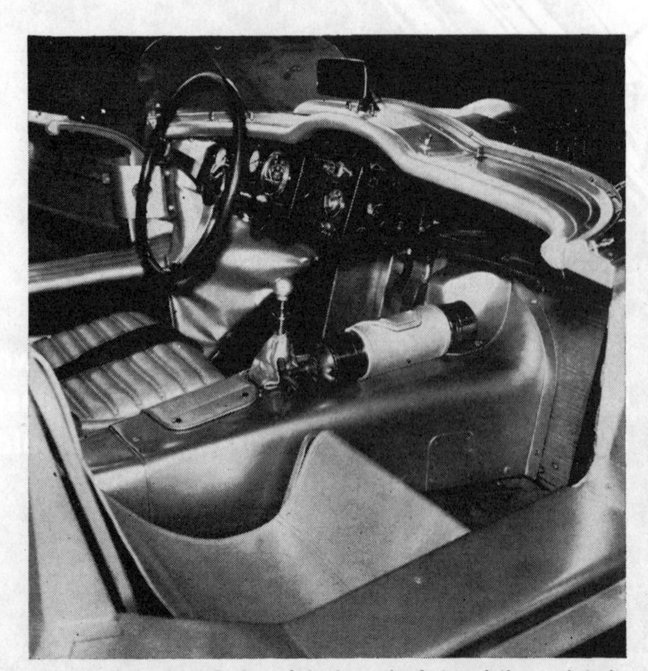

One look at the Sting Ray's cockpit shows the factory design work in its background. Sling passenger seat meets SCCA requirements. SS problem of toasted driver was solved by re-routing exhaust system and insulation.

THOMS

Sting Ray driver Dick Thompson looks happily expectant while waiting on the grid for a start. His handling of the car against smaller and more agile machinery has provided exciting — if not always successful — racing.

300 genuine horses, as installed, with 280 bhp at the clutch, at about 6200 rpm. Usual rev limit is 6500 rpm, with 6800 okay in top gear. Torque comes to a peak at 44-4500 rpm. This is done with a stock-sized Chev 283 engine, on an 11 to one compression ratio. Dean Bedford and Mitchell have resisted the temptation to "go Scarab" on displacement, feeling that the Chev only becomes less reliable in the process and more demanding of maintenance.

At Sebring the Corvette SS used the prototype aluminum cylinder heads, without valve seat inserts. These heads have been used on the Sting Ray, as have aluminum heads *with* inserts. Both work with equal high efficiency and reliability on this carefully-maintained car. Between the heads a 1957 Rochester fuel injector continues to supply fuel-air mix, set a shade on the rich side to keep the engine running cool. The normal racing fuel consumption is about 6 miles per gallon; a supplementary tank that brings the capacity to 45 gallons is used for the Road America 500.

Down below, the 30-weight Shell X100 oil is kept under control by a baffle with a little swinging trap door in the oil pan, and by a remarkable Dow-Corning anti-foam additive ("200 Fluid") which absolutely eliminates bubbles in the oil. The bypass-type oil cooler, mounted ahead of the radiator to the right, is an experimental Harrison unit designed for a truck transmission. It's sensitive to the oil viscosity (when the oil gets thin it passes more of it) in such a way that it keeps the lubricant at virtually constant temperature – 240 degrees in this case – without a thermostat. Total capacity of the oil system is 7 quarts.

REFINED ACCESSORIES

A 1960 Corvette aluminum radiator, suitably modified regarding fillers and pipes, handles the cooling job perfectly. It has to be angled well forward to fit under the low hood, which makes it possible to duct the exit air up and out in a most interesting way. Actually only a third of the air goes out through the two grilles in the hood, the rest flowing back through the engine room to keep the compartment cool. A 180-degree thermostat is used, and the aluminum water pump is driven at .7 instead of .9 engine speed in view of the consistently high revs used.

A standard Corvette clutch transmits the power, but it has to be a "good" one. The four-speed box has an aluminum case, experimental in 1957 but now standard on Corvettes, possibly by virtue of its trial by fire in the Sting Ray. At the back is a Halibrand center section termed "intermediate-size" – probably the V8-type unit after it was beefed up to double its capacity. At any rate when the SS was built Chevy bought the raw castings and machined them to take its own side plates and ring and pinion. In use on the Sting Ray, the bearings in this unit have to be changed every 1500 racing and practice miles, to be on the safe side, indicating that it's definitely overstressed in this application. Dean Bedford makes a keen point here: "The SS Corvette design was adequate for Sebring but would never have finished Le Mans."

A Powr-Lok limited-slip differential is used. Naturally a wide selection of ratios is at hand, beginning with the 3.55 that was fitted to the SS at Sebring. This isn't usable on the Sting Ray on U.S. tracks, and Bedford feels that for the Florida track he'd choose 3.70 today – the cogs they use for Bridgehampton and Meadowdale, where the car reaches about 155 mph. For Danville and Road America 3.86 gears are used, 4.20 for Cumberland and 4.39 for Marlboro. These can of course be matched with the tire sizes chosen for the 5½K x 15 Halibrand rims. Usually Firestone 6.70 x 15 is worn at the front and 7.10 x 15 at the back, though the smaller sizes can be fitted all around for tight tracks like Cumberland and Marlboro.

PROVED ITS POINT

Certainly Bill Mitchell's exciting Sting Ray has proved a strong point on behalf of the Corvette SS, with some excellent racing performances, especially in view of the time that's passed since the design was laid down. Where's the SS today? It's something of an old warhorse around Chevrolet Engineering, its very fragile magnesium body held together by rivets and baling wire. From the competition standpoint, its promise has now been amply fulfilled by its alter ego. Now we look forward to seeing more of the experience gained with these cars employed in the Corvettes of the future. *–KEL*

WHEELER

The Sting Ray body may provide design ideas for future production Corvettes. Recently car was raised in the rear to kill lifting tendency at high speed.

1962 CHEVROLET CORVETTE

Of course we tested the storming new injected Corvette, with 360 horses from 327 cubes, but this time we added the "smallest" engine with a smooth Powerglide for fun on the open road.

● Zora Duntov says that as far as he's concerned, this is the last time the Corvette's power will be increased—within the present chassis, anyway. It's a reasonable statement, since it's hard to discern a need for more poke than the booming new 327-inch Corvette V8 provides. The 360-horsepower fuel injection version more than makes good on the public relations promise of "added excitement for those wishing the ultimate in performance."

Recently we've always tested the hot fuel-injected version of the Corvette; we followed through on this with the '62 car. But in response to many requests from readers for data on a more normal go-to-market Corvette, we explored the opposite end of the range this year with the "small" (a miserable 250 horses) engine linked to the new aluminum-case Powerglide transmission. In this connection it's interesting that at Byrne Chevrolet's "Corvette Corral" in New York's Westchester County, the stock engine plus the four-speed box accounts for 80 percent of Corvette sales.

First, let's look at this new 327-cubic-inch engine, which makes official what hot-rodders have been doing to Chevys for years now. It's definitely based on the 283-inch engine design, keeping the same bore center-to-center distance of 4.40 inches while increasing the bore ⅛ inch to an even 4.0 inches. (Since the new Chevy II four and six have the same center-to-center distance, with redesign they too could presumably abide a four-inch bore.) Stroke is stretched ¼ inch from the 283's 3.0 inches. Though the 327's rod length and bearing sizes are unchanged, the rod does have a heftier shank section to take the bigger engine's higher stresses. For the same reason all 327s have the heavy-duty aluminum Moraine bearings that were used only on the hot 283s before.

Last year's big-port head, with 1.94-inch intake valves, is now used on the top three of the four Corvette engine options. Only the base engine has the 1.72-inch intakes, fed by a normal four-barrel carb to deliver 250 bhp. The next hotter engine stays with the hydraulic-lifter cam but moves to the big-port head, and is equipped with a new oversized four-barrel Carter carb. Its output is 300 bhp, and it's still available with the Powerglide box. These two engines have double head gaskets for a compression ratio of 10.5 to one (actually about 10.2 to one); for the two top engines one gasket is pulled out to raise the c.r. to a nominal 11.25 to one, actual 11.1 to one. The Duntov solid-lifter cam is also fitted, output being 340 bhp with the big four-barrel and the aforementioned 360 with injection. The famed dual-quad layout, with all its complexities, is now entirely replaced by the new big Carter four-barrel.

Rochester fuel injection goes into its sixth year of production with a major change to adapt it to the deeper breathing requirements of the big V8. Instead of the old relatively complicated cold start arrangement, the new injector has a simple choke valve in a port in the center of the intake venturi plug. When this "strangler" valve is open, the total area made available is adequate to the engine's needs; when it's closed, it has a definite choking effect although the venturi itself—necessary for metering reasons—remains open. Control of the choke is fully automatic.

As always, the fuel-injected Corvette engine is a sweetheart to drive. Power is excellent, but even more impressive is the torque curve, which is as close to being flat as any we've seen. From its 4000 rpm peak it falls off only 20 pound-feet at the extremes of peak power (6000) and 2000 rpm. In action this engine delivers such constant thrust without surges or flat spots that the feel of acceleration is deceptively docile. The figures prove that performance is much like last year's f.i. Corvette (SCI, December, 1960) up to 80, above which the new one just continues to soar. Though the time to 80 is unchanged, time to 100 is cut from 18.0 to 16.6 seconds. Quarter-mile time is improved from 15.6 to 15.0 seconds, though the terminal speed is up only one mph to 95—not bad for a stocker!

With two aboard, normal tires, no positraction and a 3.70 axle, this Corvette was highly standard. The added power is such that once again rear-axle bounce is a problem on standing starts, in spite of the torque arms above the axle, so the times to low and medium speeds can't show improvement in proportion to the new output. Though the new engine's red-line is 6300 (down 200 rpm from last year), experienced Corvette test driver Bob Clift recommended that we shift up at 5600 rpm for best results over the quarter-mile. His tests have shown that rotating inertia forces above 5600 are such that the extra revs cost more power than they produce, in terms of acceleration. Top speed would be another matter, of course.

When we jumped out of this injected bombshell into the basic Powerglide Corvette, we expected quite a comedown in performance; we were amazed at how little difference there was. Sixty in 8.8 seconds and the quarter-mile in 16.8 is, after all, not waiting around. Doing this by merely stepping down and holding on offers its own peculiar pleasure, as the solid V8 bites into higher and higher speeds with its distinctive hard, metallic wail. As a comfortable cross-country tourer, the Powerglide Corvette offers a high pleasure quotient.

During our acceleration tests of the Powerglide car it was shifting up at 4500 rpm instead of the specified 47-4800 (tach red-line is 5500), but holding it longer in the lower cog didn't change the times significantly. It makes that shift at about 58 mph, automatically under full throttle, and kickdown can be obtained at any speed up to about 47 mph. Just easing away from a standing start with light throttle, it drops into high at 14 mph. There is some creep at a standstill, though no more than is usual nowadays.

A downshift can be initiated by moving the floor-mounted control lever, but it's not the smoothest operation for braking purposes and the single low ratio isn't actually usable in the speeds where engine braking is most helpful. Thus we don't feel there's much to be gained by downshifting the box to slow down. This does give the brakes a lot of work, but even the standard Corvette binders are remarkably durable these days. For the sake of retaining maximum control, even on the highway, we feel most Corvette buyers will want one of the hand-shifted transmissions, but we'd suggest consideration of the Powerglide for a unique marriage of sports car handling and go with touring car convenience.

For 1962 there are no Limited Production Options (LPOs) for the Corvette; everything optional falls in the RPO (Regular Production Option) category. In gearboxes for example the Powerglide is RPO 313, and for 1962 there's a new RPO 685 four-speed box that's sold only with the two milder hydraulic-lifter engines.

Continuity of design is maintained for 1962 with the only grille change being that the insert is black rather than shiny as in '61.

The fuel injection model has a new choke arrangement for the 327 cubic inch mill, consisting of a flap in the venturi plug.

Beside the grille, other 1962 Corvette recognition points are fine-toothed side cut-out fillers and aluminum rocker panels.

Highway handling of stock 1962 Corvette is fine, but most road racing competitors use the 1959 stiff suspension kit.

Its ratios are 1.00, 1.51, 1.92 and 2.54, with 2.61 reverse. A specially-recommended axle ratio for use with this is the new 3.08 to one cog (RPO 203), a combination which gives an overall low gear of 7.83 to one, for blasting dramatically out of drive-ins and away from lights, yet offers a high highway top gear. Combining 3.08 with the original 2.20 would give a too-high bottom cog at 6.78 to one, so it can be said that the new box makes it possible to offer the 3.08 gear for the two quietest Corvettes.

Positraction is still available, as RPO 675, as is the RPO 687 metallic brake and quick steering kit. You can get the brake linings alone as RPO 686, and, according to CORVETTE NEWS, you can buy over the counter the parts needed to convert to quick steering for about $25. You need one adapter plate, no. 3747588, two studs, no. 3747591, and one U-bolt, no. 2066840. That price would include installation and wheel alignment, according to CORVETTE NEWS. The latter, by the way, is one real bonus to buying a Corvette. It's a top-notch publication with all the factory dope on tuning, specifications and Corvette club activities.

On the surface the 1962 Corvette is merely refined in detail. Its grille is now painted black, and the side cutouts are now fitted with fine-toothed "grilles" in place of the old triple dentures. Bright metal removed from this area has been placed low on the sides in the form of aluminum extrusions; this shouldn't be hard to remove. Under it all is the durable fiberglass body that's played such a part in keeping the Corvette's resale value at a remarkable high. The car isn't inexpensive but it's a good investment, both financially and in terms of some of the most exciting driving to be had.—*C/D*

ROAD TEST:

1962 CHEVROLET CORVETTE

Price as tested: $3887 with fuel injection and 4-speed gear box
$3897 with carburetor and automatic transmission

Manufacturer:
Chevrolet Motor Division
General Motors Building
Detroit 2, Mich.

ENGINE (4-speed)

Displacement	327 cu in 5370 cc
Dimensions	8 cyl, 4.00 in bore, 3.25 in stroke
Valve gear: pushrod-actuated overhead valves	
Compression ratio	11.25 to one
Power (SAE)	360 bhp @ 6000 rpm
Torque	352 lb-ft @ 4000 rpm
Usable range of engine speeds	700-6300 rpm
Corrected piston speed @ 6000 rpm	3615 fpm
Fuel recommended	Premium
Mileage	10-17 mpg
Range on 16.4-gallon tank	160-280 miles

DRIVE TRAIN:

Gear	Syn-chro	Ratio	Step	Overall	Mph per 1000 rpm
Rev	No	2.66	—	8.36	9.5
1st	Yes	2.20	33%	8.14	9.7
2nd	Yes	1.66	27%	6.15	12.9
3rd	Yes	1.31	31%	4.84	16.3
4th	Yes	1.00	—	3.70	21.4
5th					

Final drive ratio: 3.70 to one. (3.08 to 4.55 range available with Positraction.)

ENGINE (Automatic)

Displacement	327 cu in, 5370 cc
Dimensions	8 cyl, 4.00 in bore, 3.25 in stroke
Valve gear: pushrod-actuated overhead valves	
Compression ratio	10.5 to one
Power (SAE)	250 bhp @ 4400 rpm
Torque	350 lb-ft @ 2800 rpm
Usable range of engine speeds	1000-5500 rpm
Corrected piston speed @ 4400 rpm	2650 fpm
Fuel recommended	Premium
Mileage	13-18 mpg
Range on 16.4-gallon tank	215-295 miles

DRIVE TRAIN:

Gear	Syn-chro	Ratio	Step	Overall	Mph per 1000 rpm
Rev	Auto	1.82	—	6.09	10.6
Low	Auto	1.82	—	6.09	10.6
			82%		
Drive	Auto	1.82-1.00	—	6.09-3.36	19.4

Final drive ratio: 3.36 to one

CHASSIS:

Wheelbase	102 in
Tread	F 57 in, R 59 in
Length	177 in
Ground clearance	6.7 in
Suspension: F, ind., wishbones and coil springs, anti-roll bar; R, rigid axle and semi-elliptic leaf springs, radius rods and anti-roll bar	
Turns, lock to lock	3¾
Turning circle diameter, between curbs	L 37 ft, R 36½ ft
Tire and rim size	6.70 x 15, 15 x 5 K
Pressures recommended	Normal, F 24, R 24 psi
Brakes, type, swept area	11-inch drums, 259 sq in
Curb weight (full tank)	3048 lbs (Automatic) 3045 lbs (4-speed)
Percentage on the driving wheels	48

ACCELERATION:

Zero to	4-speed Seconds	Auto-matic Seconds
30 mph	3.0	3.9
40 mph	4.1	5.2
50 mph	5.3	6.6
60 mph	6.9	8.8
70 mph	8.7	11.4
80 mph	11.0	14.6
90 mph	13.6	18.8
100 mph	16.6	24.6
Standing ¼-mile	15.0	16.8

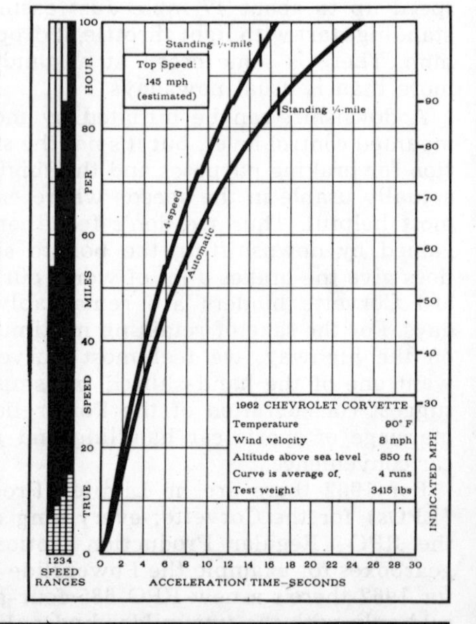

Labor Day Toll: XK-E Succumbs to Corvette

Bill Krause in the E-Type leads Paul Reinhart's Corvette through Turn 9 at Santa Barbara as the cars competed for the first time.

Veritably surrounded by Corvettes, the lone Jaguar fought hard. The E managed to finish third twice in two days of racing action.

Turn 9, again, sees Bob Dickson and Krause still in a nip and tuck battle. The E trailed a Corvette and Porsche to the finish.

● The first round of the battle American enthusiasts have been longing to see—Corvette versus Jaguar on U.S. tracks—ended in a decided victory for the flying fiberglass. The scene was Santa Barbara, California, September 2 and 3.

The flag dropped and Bill Krause whipped the E-Type roadster into the lead in Saturday's event, trailed by a gaggle of Corvettes. By the fifth lap the picture changed completely as they ran Corvette, Corvette, XK-E. Bob Bondurant racked up a 13-second lead and was notably unruffled. Behind him Krause and the E-Type were having trouble fending off a searing attack by Paul Reinhart in another Corvette, this one running two four-barrel carbs instead of fuel injection. Coming out of Turn 9, Reinhart shot past Krause and that's how they finished: Bondurant, Corvette; Reinhart, Corvette and Krause, XK-E.

By Sunday the crowd's curiosity was insatiable. Had Krause been holding back? Was the E running right? Was it set up for the circuit? Bondurant, his car being protested and found to have an aluminum flywheel, was placed in the big-bore modified class. The pack blasted off, Krause assuming an initial three-length lead over Corvettes of Bob Dickson and Reinhart. On the next lap, Reinhart went deep into Turn 9, passing both Dickson and Krause, picking up a length's lead. In the same corner on lap 3, Don Wester's Porsche Carrera passed the E-Type, leaving it fighting hard for third overall. When the checker fell, Reinhart's winning Corvette led the E by 20 seconds at an average 1.7 mph less than Bondurant's Saturday speed.

Santa Barbara's spectacle raises some interesting points in the hot issue of Jag vs. Corvette. For one thing, the 'Vettes were '61s. Based on the data in our '62 test in this issue it's likely the competition will become even more fierce. Yet it's quite likely the E had not been set up the way the British cars have been and these changes could conceivably have helped Krause. As it happened, the early lead by Krause supports acceleration data we recorded. Finally, it would be very un-Jaguar if the factory didn't have a performance ace or two up its corporate sleeve. Inconclusive? Perhaps, but it should make 1962 competition something to look forward to!

—Gordon Chittenden

Top of '63 Chevy Line: the Corvette Sting Ray

The new Corvette has two body styles, a sport coupé and a convertible, with an all-new steel frame and fully independent suspension

One glance at the new Corvette tells you that it is faster and sportier than its predecessors. And when you drive a Corvette Sting Ray, either the convertible or the fastback Sport Coupé, you find that the excitement is far more than skin-deep. Hiding independent rear suspension under its sculptured tail, the Corvette is now second to no other production sports car in road-holding and is still the most powerful.

The biggest innovation this year is that the coupé is now a separate, distinct body rather than an alternative top for a basically open roadster. As is traditional with Corvette, both the coupé and the convertible are made of fiberglass. There is good reason to believe that the new coupé was inspired by the dramatic introduction of Jaguar's XK-E coupé at the Coliseum in April, 1961. GM personnel are known to have borrowed the keys to the Jaguar and come back late the night of opening day to measure and sketch it inside and out. Just as the XK-E coupé is faster than the roadster, so is the Sting Ray Sport Coupé faster than the convertible. With open exhausts, the coupé has exceeded 160 mph, and with full street equipment it will still be able to do a genuine 155, whereas the top speed of the convertible lies between 150 and 155, according to Zora Arkus-Duntov.

More prosaically—but very important—the seating position has been tremendously improved over previous Corvettes. The seats are farther back and so many adjustment possibilities are provided that any driver should be able to find a comfortable position. Straight-arm driving is at last possible, and visibility over the sloping hood is very good. From behind the wheel, the car looks and feels smaller than the "old" Corvette, and it is.

The wheelbase has been reduced from 102 to 98 inches, and the front track is down from 57 inches to 56.3, while the rear track has been reduced from 59 inches to 57. All of this makes it more nimble and more easily steered with precision.

When it was boosted to 327 bhp last year, it became obvious that the Corvette could not again be improved by merely adding power. With the chassis thus at its limit, it was decided to make improvements where they were needed most: in the frame and suspension. Zora Arkus-Duntov, H. F. Barr, and E. J. Premo of Chevrolet Engineering got together to build an all-new chassis, drawing extensively on their experience with the Corvette SS of 1957, the Sting Ray, and the CERV-1.

The geometry of the independent rear suspension has been directly derived from the open-wheel racing car protoype, CERV-1. There was no room on the '63 Corvette for coil springs as used on the CERV-1, so Duntov chose a transverse leaf spring and put it behind and below the differential housing.

Unsplined half-shafts with two universal joints each function as upper wishbones. Underneath each is a simple lateral control rod; driving and braking thrust are taken up by stamped-steel radius arms. The differential housing is bolted to the frame both front and rear.

Compared with the rigid axle of the 1962 Corvette, the new model has a lower rear roll center (8.13 inches above ground level against 9.0). This is the same as on the CERV-1, while

Two new Corvette models acceler

the de Dion-suspended SS and St
Ray had roll centers 8.0 inches ab
the ground.

In their normal position, the r
wheels have 1.5° negative camb
In full compression, this is increa
to 5.5° negative camber, and full
bound gives a positive camber of
Wheel travel is 3.15 inches on co
pression and 4.0 inches on rebou
The design originally called for 3
inches of travel on compression,
GM Styling's fender line force
slight reduction. The front whe
rise the full 3.75 inches on compr
sion and fall four inches on rebou
with very small camber variatio
3.18° negative on full compressi
1.73° on full rebound, with a norr
setting 0.5° negative. Spring rates
the wheel are 105 lbs/inch at fr
and 125 lbs/inch at rear, giving
higher static deflection at the fro

The Corvette Sting Ray has a m
nearly horizontal roll axis than
1962 model, as the front roll cen
has been raised 3.4 inches, while
rear has been lowered 1.13 inch
This has increased the roll stiffn
and contributes to more balanced
actions in the car on turns.

Most important advance in
chassis design lies in its geomet
The rear suspension also bri
about a sizable reduction in u
sprung weight, for the unspru
parts in the 1963 rear suspension a

ay. Improvement in cornering power over the previous Corvettes is striking. The styling reflects Bill Mitchell's Sting Ray of 1959.

to about 210 pounds, while the r axle, wheels and unsprung parts the leaf springs and radius rods of 1962 Corvette weigh 350.

With the switch to independent rear suspension, it was natural Chevrolet to design a brand-new me. They knew that compared h the X-braced, boxed-in channel el frame then used, a simple steel e frame (as on the production-del Ferrari) has many advanes. It is light and rigid, and offers, h proper design, correct attachnt points for the body work. The evrolet engineers designed a similar tubular frame and a prototype s made. Then, because tubing ts so much, built-up box sections re substituted one place after aner. The frame ended up with no es at all by the time it was resed to production—heavier than optimistic prototype, but more nomical to produce.

Knowing so much more about rts cars than they did in 1953, y put the frame members outrd of the seats instead of directly der them. As well as giving a more nfortable seating position, this mits a much lower roof. The total ght of the coupé is only 49½ inch- (the top of just the windshield of 1962 model stood 52 inches off ground). The body shape has

Third basic body shell for the Chevrolet Corvette is the most radical and also the most practical. Today's Corvette Sting Ray is much closer to thoroughbred European sports cars than to the original Corvette two-seater introduced in 1953.

The sleek Corvette Sting Ray Sport Coupé is only a two-seater but can hold vast amounts of luggage behind the seats. Its tail accommodates both spare tire and fuel tank. Shape of the body has been carefully checked out in wind-tunnel tests.

Corvette's new chassis carries the engine and complete drive line offset one inch to the right of the longitudinal center plane to increase the driver's legroom.

been carefully tested in wind tunnels both at full scale and with 3/8-scale models. It was decided not to incorporate a belly pan, since a flat bottom would only raise the maximum speed by about two mph and the added weight would hurt the acceleration.

The center of gravity has been lowered from 19 inches in the 1962 model to 17.5 inches and moved slightly rearward. The weight distribution with the 20-gallon tank full is almost exactly 50/50, the rear wheels carrying perhaps 14 pounds more of the 3,012 total. The 1962 model had a 52/48 weight distribution, with a curb weight of 3,048 pounds on a full 16.4-gallon tank.

Though the 34-pound weight reduction seems surprisingly small, there are several reasons for this. Heavier gauge metal is being used in the frame and body reinforcements, and the exhaust system alone, a new design adopted for its longer life, weighs 80 pounds. Then there are various luxury items such as the concealed headlights with their electric motors and hinge mechanism.

One of the necessary luxuries, and therefore one of the most appreciated, is the introduction of fully adjustable seats. The backrest rake is infinitely variable, and the seats have a four-inch fore-and-aft travel. The seat height above the floor has three positions, with a difference of 1.24 inches between upper and lower. In addition, there is a three-inch adjustment on the steering column to get the wheel closer or farther away. The pedals are extremely well placed. The accelerator is close to the tunnel, letting the foot rest against its side, and there is plenty of room left of the clutch pedal for the driver to brace himself. The importance of these adjustments cannot be overestimated, and it is highly encouraging that Chevrolet has adopted them all.

The now-usual heavy-duty (i.e., racing) kit includes stiffer front and rear springs and front anti-roll bar, cast-aluminum wheels with knock-off hubs, an extra-large (36½ gallons) gas tank, finned aluminum brakes, sintered metallic brake linings, and dual master cylinders separating the front and rear brakes. Required by all who plan to race the Sting Ray in serious competition, the kit will be for the coupé only.

Evidently Chevrolet wishes to keep the convertible out of racing. One reason is surely the obviously stronger body structure of the coupé. Not only does the roof itself add rigidity, but there are roll-bar-like steel reinforcements on both sides that extend from the frame and meet in the roof.

A Saginaw mechanical (recirculating ball) steering gear is standard and incorporates a lovely refinement: the steering arms are all made with two tie rod attachment holes, so that the driver can easily adjust the steering ratio from 19.6-to-one to 17-to-one. The slow steering gives 3.4 turns of the wheel lock to lock, and the quick ratio only 2.9.

Power steering is optional with the 250-bhp and 300-bhp engines only, and uses the quick steering ratio. There is a built-in steering damper on all linkages (except the heavy duty steering) but it is set to give some feedback even with the power steering in order to increase the "feel" of the road.

Even the "slow" manual steering is not very slow, although right-hand street corners cannot be negotiated without changing the grip on the wheel rim. With the quick steering, the hands need not be moved from their normal location.

Thanks to the improved geometry and the rearward relocation of the center of gravity for 1963, steering effort is very low, and the car inspires a high degree of confidence which continues to grow with closer acquaintance.

Cornering behavior is extremely stable, and the car has the added advantage of a power reserve so that a high-speed drift can be entered and maintained even by moderately skilled drivers. In a balanced drift the tail hangs out, but very little correction is called for, as throttle steering seems the natural way of aiding the car round the curve. This must not be misunderstood to mean that the car is dependent on engine power to pull it out of a turn. The Corvette can actually be flung around tight corners in neutral, without loss of adhesion, but naturally it loses speed in the process.

General Motors is still unwilling to go to disc brakes, mainly for cost reasons, although experiments with disc brakes are being conducted con-

New independent rear suspension carries the transverse leaf spring in such a way that its mass counts as sprung weight. Unsprung weight at the rear is 210 pounds.

tinually. While it is interesting that the original Sting Ray has recently been fitted with Dunlop discs, the production Corvette Sting Ray has drum brakes with 18% more friction area than on the 1962 Corvettes. The new standard brakes are adjusted automatically when the car is braked while backing. The optional brake system with sintered metallic linings and twin master cylinders have a different automatic adjustment, which operates on forward motion of the car. The tires are the same size as before, 6.70 x 15, with 5.5-inch wide-base steel bolt-on wheels and bright-metal hub caps. Cast-aluminum wheels with knock-off hubs are available as an option.

All engine options are based on the now-familiar 327-cubic-inch V-8 with four-inch bore and 3.25-inch stroke. With a 10.5-to-one compression ratio and a small four-barrel carburetor it develops 250 bhp at 4,400 rpm. With the same compression and a large four-barrel carburetor, output is raised to 300 bhp at 5,000 rpm. A compression increase to 11.25 to one together with the large four-barrel gives 340 bhp at 6,000 rpm. This latter engine, with fuel injection instead of carburetion, produces 360 bhp at 6,000 rpm.

Steel reinforcements for the fiberglass body on the sport coupé are carefully contoured around the passenger compartment, and are an important safety feature.

Powerglide is optional with the 250-bhp and 300-bhp versions, while a manually controlled three-speed gearbox is standard. The all-synchromesh four-speed transmission is optional with all engines.

The standard final drive ratio is 3.36 to one, though 3.70 to one is available without Positraction with the two most powerful engines. Positraction non-slip differentials offer a very wide choice of ratios: 3.08, 3.36, 3.55, 3.70, 4.11 and 4.56 to one.

The Powerglide option is the same as on the 1962 model, with a floor-mounted lever moving in the familiar P-R-N-D-L quadrant. Low and reverse ratios are 1.76 to one, and the maximum torque conversion ratio at stall is 2.10 to one.

The four-speed transmission has the same close ratios as last year: 2.20 to one in first, 1.66 in second, 1.31 in third and direct in top. The three-speed gearbox has a 2.47-to-one low gear, a 1.53-to-one second and a direct third. This transmission has synchromesh on second and third.

Considerable effort has been made to enrich the interior of the Corvette. The 1963 models have all-vinyl interiors, and Chevrolet plans to offer genuine leather upholstery later in the year. Seat belts are standard.

The floor covering is a deep twist carpet, which is also used in the luggage space behind the passenger compartment. The headlining for the sport coupé is molded vinyl-coated fiberglass, which is hard to stain and easy to clean.

The center console and the dashboard hood are trimmed with a padded vinyl cover. Above the glove compartment is a grab rail.

GM Styling has found a way to combine ornamentation with function in several instances. The large Corvette emblem on the rear deck conceals a large centrally located fuel filler cap. The flush-fitting headlights recalling those of the 1936 Cord and the '42 De Soto not only improve the aerodynamics but in daytime motoring also protect the lenses from dirt and stones.

Much attention has been paid to ease of entry and exit. The gently curved windshield has straight corner pillars, and the coupé doors extend into the roof panel to add entrance height. The deep twin windows give a good rear view.

The concealed quadruple headlights are mounted on hidden hinges and operated by electric motors coupled to light switch.

Air evacuation from the engine takes place under the car. The functional grilles on the hood of the original Sting Ray become merely dummies on the 1963 Corvette, because they would draw air and fumes into the heater-ventilator air intake on the cowl. If it turns out that aerodynamic lift at racing speeds is a problem, making them genuine with a hacksaw would be easy.

Chevrolet was first to offer a genuine American sports car. In its first stage it was strictly a boulevard job. When the T-Bird stole that market, Duntov had the delightful task of making a real racer out of it—and succeeded in producing a car which has earned a respected place on the track. In a pragmatic way, he did it with brute power and a minimum of subtlety. Now the Corvette enters its third era. It is still tremendously powerful (less than 10 pounds per bhp), but now it has suspension to match its speed potential.

Prices of the two Corvette Sting Ray models have not yet been announced, but Chevrolet spokesmen have indicated that no appreciable increase is expected. If this is true, they offer fantastic value for money, whether you want to race or drive fast over long distances in comfort, or merely need a smart-looking car to use around town. C/D

New instrument panel has all the vital instruments grouped directly in front of the driver, with the clock on the right.

Mr. Corvette and His Cars

By Jan P. Norbye

Zora Arkus-Duntov is so firmly identified with Corvettes they could bear his name

With the original Sting Ray and the Corvette Sting Ray sport coupé in the background, C/D's tech editor (left) discusses Corvette development with Duntov.

After ten years of Corvette production, with constant improvement year by year, it is a pleasure to report that the 1963 car (October C/D) represents a major advance on its predecessors. It is the most radically redesigned Corvette since the inception of that model name, and many of the changes were derived from a number of experimental Chevrolets, some of which were raced extensively. Great credit is due to Chevrolet's engineering development program and to the one man who sparked that program and made the new Corvette possible—Zora Arkus-Duntov, Mr. Corvette.

Paradoxically, Duntov was not connected with the design of the original Corvette. He joined Chevrolet Engineering when Maurice Olley, the suspension expert who served as a consultant on many GM experimental programs, was engaged on "Project Opel," the open two-seater which became the first Corvette. Duntov was later appointed Director of High Performance Vehicle Design and Development, with responsibility for the technical development of the Corvette as his primary duty, but also for special engine and chassis projects on other Chevrolet models.

The industry knew there was a market for an American sports car as early as 1950, and interest was sparked by such efforts as the Willys Jeepster, the fiberglass-bodied Kaiser-Darrin two-seater and the Lincoln-based Muntz Jet. While Olley was on "Project Opel," rumors reached GM that Ford was working on a similar car to be called the Thunderbird. The Corvette prototype was built in near-record time, by using standard components of proved reliability wherever possible, and America's first postwar sports car was born.

A wheelbase of 102 inches was chosen because the successful Jaguar XK-120 had that wheelbase, and several other dimensions in the first Corvette, such as seating, were deliberately identical to those of the Jaguar.

The Corvette frame was a cross-braced channel-section structure, and the engine was a three-carburetor version of the familiar Chevrolet Six tuned to develop 150 bhp at 4,200 rpm. The wishbones-and-coil spring front suspension used many standard parts, but had a heavier anti-roll bar, and the rigid rear axle was carried on semi-elliptic leaf springs.

The prototype had a fiberglass body, but a steel body was to be used in production. However, a pilot series was made with fiberglass bodies, and on the basis of their trouble-free operation it was decided to drop the steel

The original Corvette of 1953 had a six-cylinder single-carb engine, Powerglide automatic transmission and fiberglass body.

Chevrolet built the original Corvair as a GM Motorama dream car in 1955, using a Corvette chassis for this fastback coupé.

The body was restyled for 1956, the V-8 engine became available and then its competition potential first became apparent.

The 1960 Corvette was available with a rich choice of racing equipment, based on experience with the Corvette SS of 1957.

body. The least typically sports-car feature of the first Corvette was the automatic transmission, a modified Powerglide. It was not until 1955 that the three-speed close-ratio stick shift became available. This was followed by the excellent four-speed all-synchromesh unit two years later.

Zora Duntov spent most of 1956 perfecting Chevrolet's Rochester port-type fuel injection, and was also charged with setting up a factory team for Sebring the following year, with experimental Corvettes. The result was the Corvette SS, a most promising design which was killed by the AMA resolution on racing (and speed and performance advertising) after a single, not very successful racing appearance.

Its engine was a tuned 283-cube Chevrolet V-8 with aluminum cylinder heads and Rochester constant-flow fuel injection, developing 310 bhp. A close-ratio four-speed gearbox was coupled to a quick-change Halibrand rear end suspended in the specially made space frame. Rear suspension was by a de Dion tube, four radius rods, and sharply splayed coil springs with concentric telescopic shock absorbers. Inboard drum brakes were used at the rear, normal drums mounted on the front wheels.

In addition to the team cars, an extra practice car had been built, which was to have its own history. It was transformed at GM Styling, where it was furnished with a new body designed by Bill Mitchell (and renamed the Sting Ray). Later it received a series of chassis improvements. In cooperation with Duntov and Chevrolet Engineering, it was prepared for Dick Thompson to race. In contrast to the Corvette SS, it did very well, and after test-driving it recently in its latest form, with a 327-cube V-8, Dunlop disc brakes and a locked differential, we feel that it would be competitive in big-league sports-car racing today.

The Corvette SS and the Sting Ray have had a considerable influence on the design of the 1963 Corvettes, although the de Dion rear end and the tubular space frame have not been carried over to the new model. The rear suspension of the Corvette Sting Ray is based on that of another Chevrolet experimental car—the open-wheel single-seater designated CERV-1. This car, of course, was also a creation of Zora Arkus-Duntov.

With only the engine and transmission remaining from the standard Corvette, the 1963 models can be called all-new, yet they have many well-proved components. Not many manufacturers are in a position to achieve such combinations, and certainly 10 years ago Chevrolet knew less about sports cars and racing than,

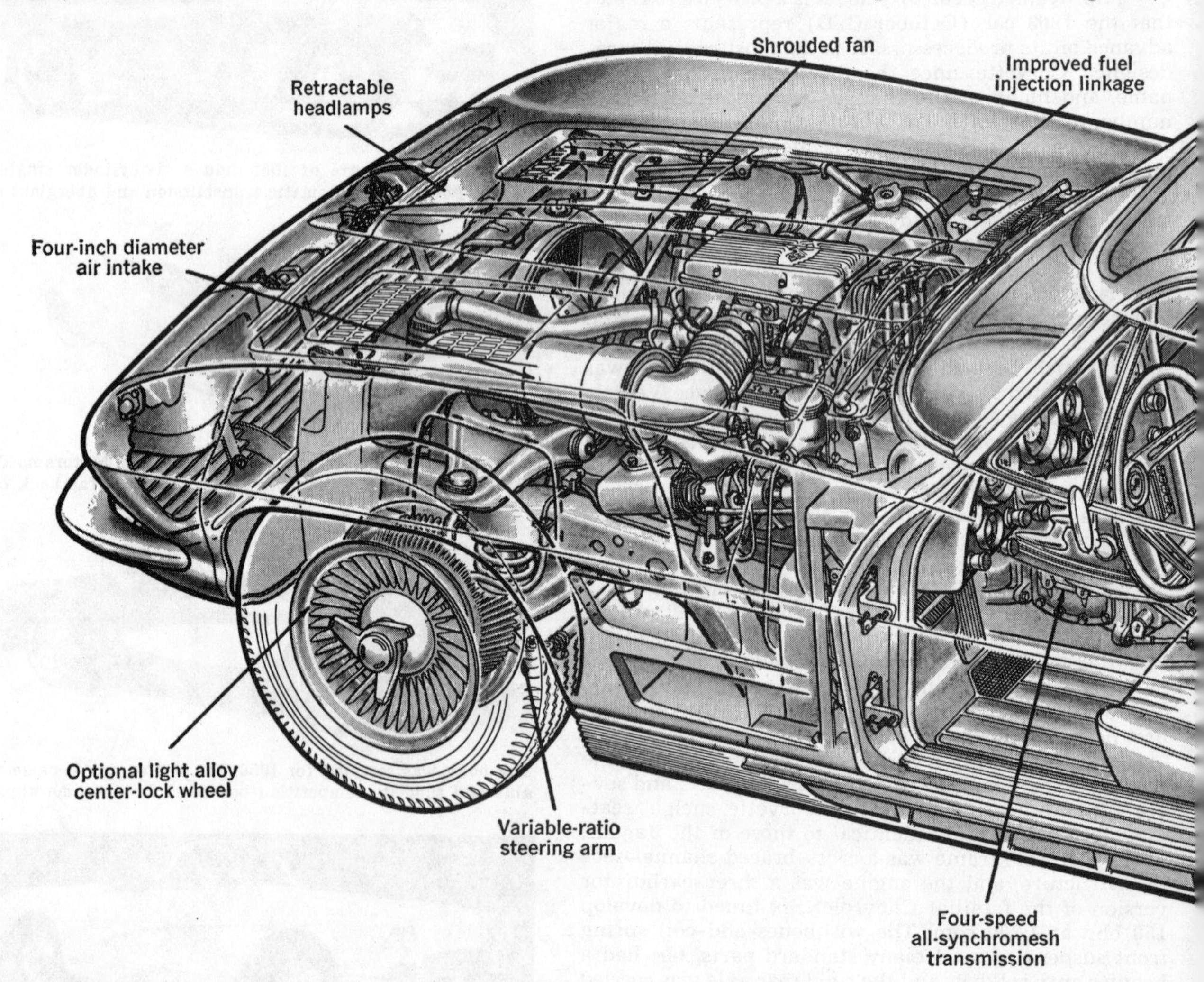

for instance, Chrysler. The long-term research and development instituted by Duntov in his first years at Chevrolet is paying off handsomely today.

To return to the CERV-1, its development differed from both the SS and the Sting Ray. In 1958, Duntov was seriously thinking of redesigning the Corvette as a rear-engined car. There were various reasons why the rear-engined two-seater was never built. Surprisingly, the most important single cause for the project being abandoned was the poor view from the driver's seat, caused by a low seating position and big front wheels. But the rear-engined ideas had matured by late 1960, when the single-seater was first shown publicly. In this design, a fully independent rear suspension had been judged superior to a de Dion layout, and the resultant system has been copied on the Corvette Sting Ray (but a transverse leaf spring has replaced the splayed coils of the CERV-1). One reason for the change in the springs lies in the fact that the 1963 Corvette is a much heavier car. If coil springs were to be used, they would have to be of larger diameter. Problems of finding space would then inevitably present themselves, especially with the production car's outboard brakes. It would have been possible to relocate the springs, but the leaf spring provides a fully acceptable solution.

Duntov's experimental cars have placed Chevrolet

CORVETTE STING RAY SPORT COUPÉ

ENGINE:

Displacement	327 cu in, 5,370 cc
Dimensions	8 cyl, 4.00-in bore, 3.25-in stroke
Valve gear	Pushrod-operated overhead valves
Compression ratio	11.25 to one
Power (SAE)	360 bhp @ 6,000 rpm
Torque	352 lb-ft @ 4,000 rpm
Carburetion	Rochester fuel injection

DRIVE TRAIN:

Clutch .... Borg & Beck 10-in single dry plate

Gear	Synchro	Ratio	Step	Overall	Mph per 1,000 rpm
Rev	No	2.26	—	—.36	—9.5
1st	Yes	2.20	33%	8.14	9.7
2nd	Yes	1.64	27%	6.15	12.9
3rd	Yes	1.31	51%	4.84	16.3
4th	Yes	1.00	—	3.70	21.4

Final drive ratio: 3.70 to one (3.08, 3.36, 3.55, 3.70, 4.11 and 4.56 available with Positraction).

CHASSIS:

Wheelbase	98 in
Track	F 56.3 in, R 57.0 in
Length	175.3 in
Tire size	6.70 x 15 (4-ply)
Curb weight	3,015 lbs

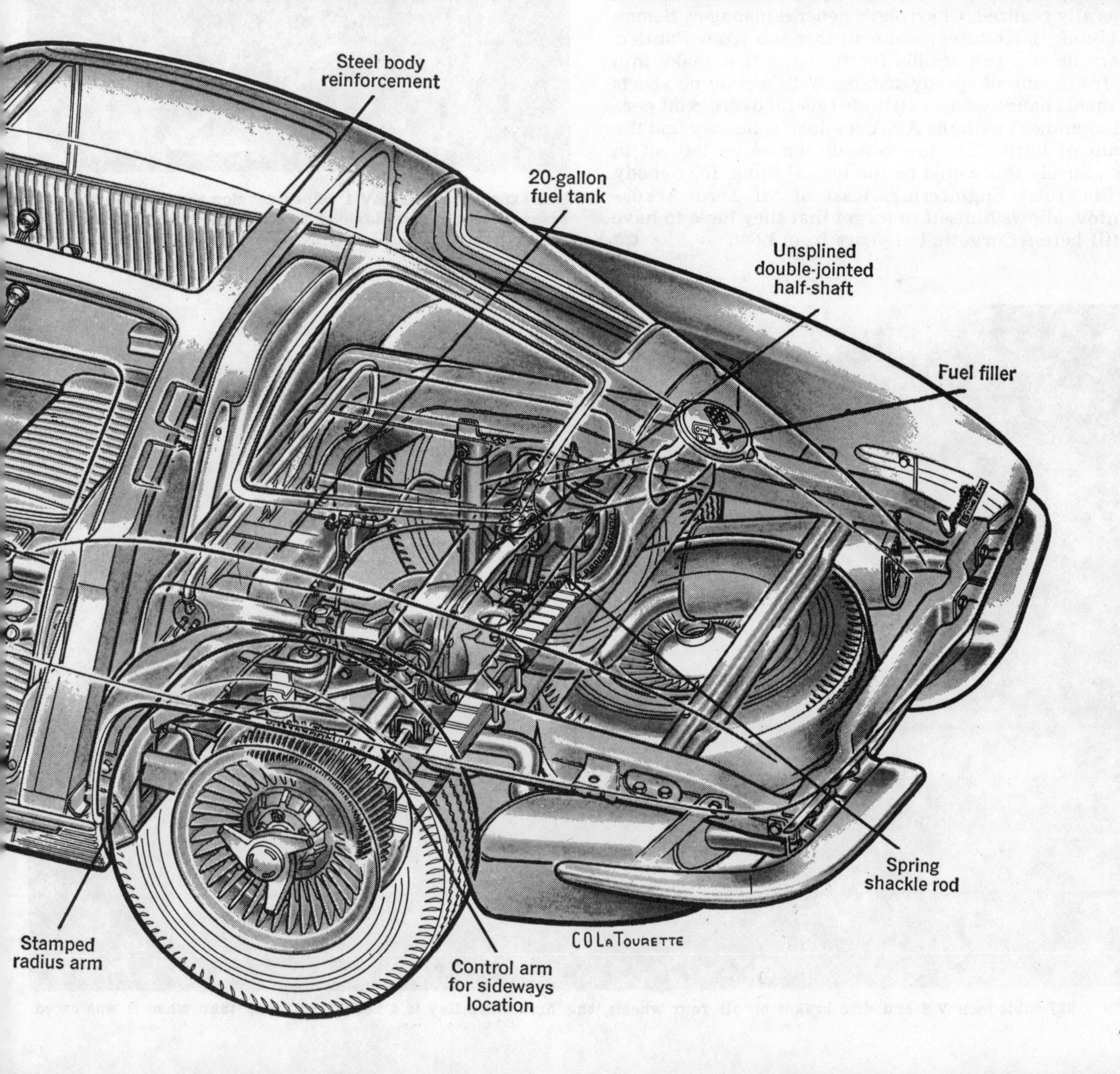

Engineering in the enviable position of rear-suspension pioneers in this country, and one might think that Duntov had long been known as an expert on suspension geometry. But before he joined Chevrolet, he was best known for his special camshafts and the Ardun overhead-valve conversions for flat-head Ford and Mercury V-8s. For several years he was the associate of Sydney H. Allard, who manufactured the Allard sports car in London, and Duntov even partnered Sydney Allard in racing, notably at Le Mans. He already had considerable racing experience with Porsches in Europe, which may have had something to do with his interest in rear-engined cars.

Duntov remains non-committal about his early past and future racing plans, both for himself and for the new Corvettes. As pointed out in last month's C/D, only the sport coupé will be available with heavy-duty (i.e., racing) options. We hope that Chevrolet has plans to race the car officially, a thought which brings only a smile to Duntov's lips, but apparently no decision has been made yet.

Perhaps optimism on this point is more realistic than generally realized. Chevrolet's general manager, Semon E. (Bunky) Knudsen, came to this job from Pontiac, where he was responsible for bringing that make into the front rank of speedy sedans. With a genuine sports car on his hands, and an attitude several degrees off perfect alignment with the AMA resolution, he may find the means of letting Duntov take up where he left off in 1957. Surely this would be the logical thing, for nobody at Chevrolet Engineering, least of all Zora Arkus-Duntov, allows himself to forget that they have to have a still better Corvette ten years from now. C/D

The experimental CERV-1 helped develop the independent rear suspension that was adapted to the 1963 Corvette Sting Ray, with alterations caused only by styling and space requirements.

With a 327-cubic-inch V-8 and disc brakes on all four wheels, the first Sting Ray is a better car today than when it was raced.

The new Corvette Sting Ray has won the coveted CAR LIFE AWARD FOR ENGINEERING EXCELLENCE

Forgive our lack of modesty here, but we agree one hundred per cent with the editors of CAR LIFE. They think the new three-link independent rear suspension gives the car handling that's far and away the best thing ever to come from Detroit. So do we. They think the performance is on a par with any production sports car ever built. So do we and, we might add, so will you. Unfortunately, not everyone has had a chance to drive one of the new ones yet, because demand has exceeded production, but when your chance comes, you won't believe it! You've never driven a sports car that rides so well, yet handles so beautifully in the bargain. You've never sat in a car that'll turn so many heads and cause so much comment among the less fortunate drivers you pass. This car is a winner! And you'll share CAR LIFE'S enthusiasm by the time you've hit forty miles per hour and second gear! . . . Chevrolet Division of General Motors, Detroit 2, Michigan.

NEW CORVETTE STING RAY BY CHEVROLET

Waiting lists of great length and duration for the Corvette Sting Ray at all Chevrolet dealers' are the best proof of the public's acceptance of the new model. We hailed the car's technical advances with great enthusiasm (*Oct. C/D*) after our brief test drives last fall.

Now it's time for an exhaustive report on America's leading grand touring car (which many drivers think of only as a sports car). We chose the 300-bhp version of the coupé, because it seems to enjoy some market preference over models equipped with the 250-, 340-, or fuel-injection 360-bhp engines.

However, the key to the personality of the Corvette Sting Ray lies neither in the power available nor in the revised styling, but in the chassis. Up to now the Corvette has been struggling to rise above a large number of stock components, notably in the suspension, where their presence created all kinds of problems that required extensive modifications for any competition use beyond normal road rallies. The new all-independent suspension has completely transformed the Corvette in terms of traction and cornering power, but it still has some faults. The standard setup on the test car seemed a bit more suitable for race tracks than for fast back-road motoring. A rigid front anti-roll bar in combination with a relatively stiff transverse leaf spring in the rear reduces the resilience and independence of the suspension of each wheel with the result that even on mildly rough surfaces the car does not feel perfectly stable. On bumpy turns it's at its worst, veering freely from one course to another, making high-frequency corrections s.o.p., but on a smooth surface it comes incredibly close to perfection. Cornering stability under conditions permitting minimal wheel deflections is remarkable, and an initial feeling of pleasant surprise

At long last America has a formidable weapon to challenge

CAR and DRIVER ROAD RESEARCH REPORT:

Corvette Sting Ray

rises to sheer astonishment when one discovers that the car can be taken off the predetermined line with ease and still complete the turn in perfect balance.

There is some understeer but the car has such a tremendous power surplus, even with the next-to-bottom engine option, that the tail can be slung out almost any old time, and after a while throttle steering seems the natural way of aiding the car around a curve. This is so easy to do that a newcomer to the car can master it in half an hour of fast driving.

Given surface roughness, the rear end becomes skittish. We experienced this with a full tank as well as one almost empty, indicating that normal loads don't appreciably affect its behavior in this respect.

One of our test cars had the new Saginaw power steering, three turns lock to lock with enough road feel to satisfy the most critical tester and observer, while eliminating all difficulties of parking and maneuvering in tight spaces. We also tested a car with manual steering, and found it so light in comparison with previous Corvettes that there can be no conceivable need for power assistance. While the power system is every bit as good as those used by Rover and Mercedes-Benz in terms of feedback and road feel, it seems strange that Chevrolet should get around to introducing it when there is no longer any need for it. The three-spoke wheel is steeply raked (15° 23') as on previous Corvettes, and its relatively thin rim offers a good grip. The entire semi-circle between nine and three o'clock is free of spoke attachments, providing a clean hold for any but the most eccentric drivers. The steering column has a three-inch adjustment for length but our test drivers all kept the wheel in its foremost (bottom) position while making the most of seat-adjustment possibilities. There

Europe's fastest grand touring cars on their home ground

CORVETTE STING RAY

are four inches of fore-and-aft travel but backrest angle is variable only by setting screws at its floor abutments. In addition, there are three seat-height positions with a total span of 1.24 inches.

The result is a range of adjustment adequate to let our test drivers (ranging in height from five-seven to six-four) find a nearly ideal seating position. Maximum effective leg room (to the accelerator) is 43.7 inches and the maximum vertical height from the seat to the headlining is 33 inches. In view of the over-all height of only 49.8 inches, this is a good example of the care that has gone into designing the living quarters of the new Corvette Sting Ray.

As the engine and drive train are offset one inch to the right to provide wider leg room for the driver, he sits facing exactly in the direction he is going, with the pedals straight in front of him. The accelerator is nicely angled for normally disposed feet, but the clutch pedal has a rather excessive travel. With standard adjustment, you cannot release it without taking your heel off the floor, causing a bit of annoyance in traffic.

Instead of a fly-off handbrake, the Corvette has a T-handle under the instrument panel labeled "Parking Brake"—one of the few features of the new model which reminds you of its relationship with Chevrolet's mass-produced sedans.

Compared with previous Corvettes, the Sting Ray is improved in almost every imaginable respect: performance, handling, ride comfort, habitability and trunk space. The trunk is only accessible from inside the car, however, since the tail is full of fuel tank and spare wheel, but the storage space behind the seats is even larger than outside dimensions indicate. A third person, sitting sideways, may come along for short rides, but will soon feel cramped from lack of headroom. An occasional extra passenger will actually be better off sitting on the console between the seats and sharing legroom with the shotgun rider.

Having driven the Corvette Sport Coupé in all kinds of weather conditions, we found the heater and defroster units eminently satisfactory. The heater fan has three speeds, and air entry is variable by a push-pull control. Warm-up is not extremely rapid but seems to be faster than average. The body proved absolutely draft-proof and water-tight.

We liked the ball-shaped interior door handles but were not convinced of the advantages of the wheel-type door lock buttons. A minor complaint is the location of the window winders, as you cannot set your knee against the door panel for bracing on a sharp turn without coming in contact with the window handle.

Brakes have long been a sore point with Corvettes, and further advance has now been made without taking the full step of going to disc brakes (which the car really deserves). The Delco-Moraine power brakes have 11-inch steel drums cast into the wheel rims, with 58.8% of the braking force being directed to the front wheels. Sintered iron brake linings are optional and will certainly be found necessary for anyone planning to race, as fade is easily provoked with the standard linings, although the cooling-off period required to restore full efficiency is very short.

Chevrolet is prepared for a fair-sized demand for special performance parts, but has restricted their application to the structurally stronger Sport Coupé. The sintered-iron heavy-duty brake system also includes vented backing plates and air scoops and a dual-circuit master cylinder. There is a heavy-duty anti-roll bar, heavy-duty front and rear shock absorbers, aluminum wheels with knock-off hubs, and a 36-gallon fuel tank. The brake mechanism, in contrast to that fitted as standard, automatically adjusts the brakes when applied during *forward* motion. To be ordered, this special performance kit (RPO ZO6) also requires the 360-bhp engine, the four-speed Warner T-10 gearbox and a Positraction limited-slip differential.

Race preparation of the 327-cubic-inch Corvette engine has been thoroughly treated by Bill Thomas in an article for the *Corvette News* (Volume 5 No. 3), a GM publication invaluable to both the active Corvette competitor and his "civilian" counterpart. For information, readers are advised to write to *Corvette News,* 205 GM Building, Detroit 2, Michigan.

For all kinds of non-competitive driving, the 300-

Luggage space is surprisingly roomy but central window partition ruins rear view.

A ventilated fuel filler cap is reached through lid. Gear positions are labeled.

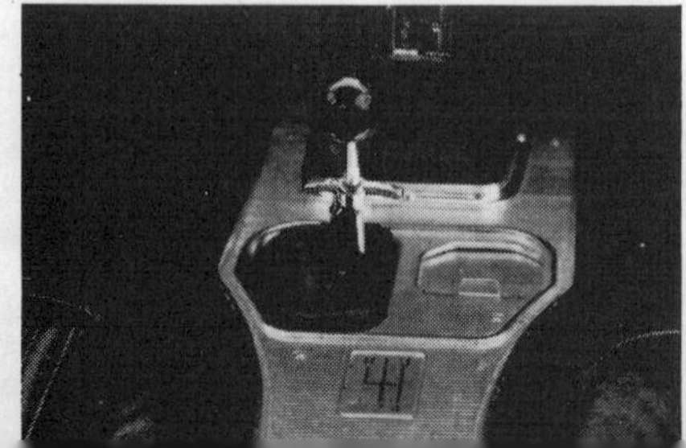

Body was wind-tunnel tested but many

bhp version gives more than ample performance for anyone, with our average standing-quarter-mile time at 14.4 seconds. This was achieved with the "street" gearbox and an axle ratio which limits top speed to about 118 mph, a combination which results in extreme top-gear flexibility as well. Top-gear starts from standstill to limit wheelspin present no problem with regard to stalling, but detonations were inevitable.

Fiberglass bodies usually have peculiar noises all their own but the Corvette was remarkably quiet, no doubt due to the steel reinforcement surrounding the entire passenger compartment. The car is also notable for low wind noise and high directional stability. Engine noise is largely dependent on the throttle opening —it will respond with a roar to a wiggle of the toe if you're wearing light shoes, and this holds true within an extremely broad speed range. Top-gear acceleration from 50 to 80 is impressive indeed, both in sound and abdominal effects.

In this connection, the gear lever has a set of speeds at which it vibrates and generates a high-pitched rattle (this is in the lever itself and not in the reverse catch), and there are intermittent peculiar noises from the clock, probably when it rewinds itself.

The now-familiar Warner T-10 gearbox has faultless synchromesh and when fully broken in can be as light as cutting butter. One interesting aspect of its operation is the fact that the owner's handbook specifies double-clutching for down-shifts.

We are in complete agreement with this recommendation, over which there has been some controversy. Some people feel that double-clutching will wear out the synchromesh. This can be true only if on down-shifts the engine is accelerated so much that the synchromesh has to work harder than it would with a single-clutch change, a situation which does not seem to occur very often.

While we agree that the Buick Riviera, for example, is the kind of car where automatic transmission has a function, we cannot see its place in the Corvette and our testing was done exclusively on a pair of manual-shift cars, one with power steering and one without, neither with Positraction limited-slip differential, which perhaps should be standard equipment on this car.

As the majority of new Corvettes are built with four-speed transmissions, it is hard to understand why the three-speed remains listed as standard equipment. We can see no reason for even continuing to offer it, and recommend that both the Powerglide and the three-speed manual gearbox be dropped. This would let Chevrolet standardize the wide-ratio four-speed transmission throughout and make the close-ratio version optional for the 340- and 360-bhp models.

Our testers preferred the car with the fewest automatic "aids," and probably most of our readers will, too. That keen drivers prefer manual controls is not baffling at all—except possibly to advanced research personnel who forget that nowhere else can they get an effective 180-pound corrective computer which can be produced at low cost by unskilled labor.

Vastly more practical than any previous Corvette, the Sting Ray Sport Coupé appeals to a new segment of buyers who would not be interested in a convertible, and production schedules at the Saint Louis assembly plant have been doubled from the 1962 model's. As an American car it is unique, and it stands out from its European counterparts as having in no way copied them but arrived at the same goal along a different route. Zora Arkus-Duntov summed it up this way: "For the first time I now have a Corvette I can be proud to drive in Europe." We understand his feelings and are happy to agree that the Sting Ray is a fine showpiece for the American auto industry, especially since it is produced at a substantially lower price than any foreign sports or GT car of comparable performance. **C/D**

object to superfluous decoration by emblems and dummy vents.

Directional and parking lights are part of bumper design but the retractable headlamps are concealed for daytime driving.

The Corvette is perhaps best looking from behind, and this is a view that drivers of other cars will soon become used to.

Road Research Report
Chevrolet Corvette Sting Ray Sport Coupe

Manufacturer:	Chevrolet Motor Division General Motors Corporation Detroit 2, Michigan
Number of U.S. dealers:	7,000 (approximately)
Planned annual production:	16,000

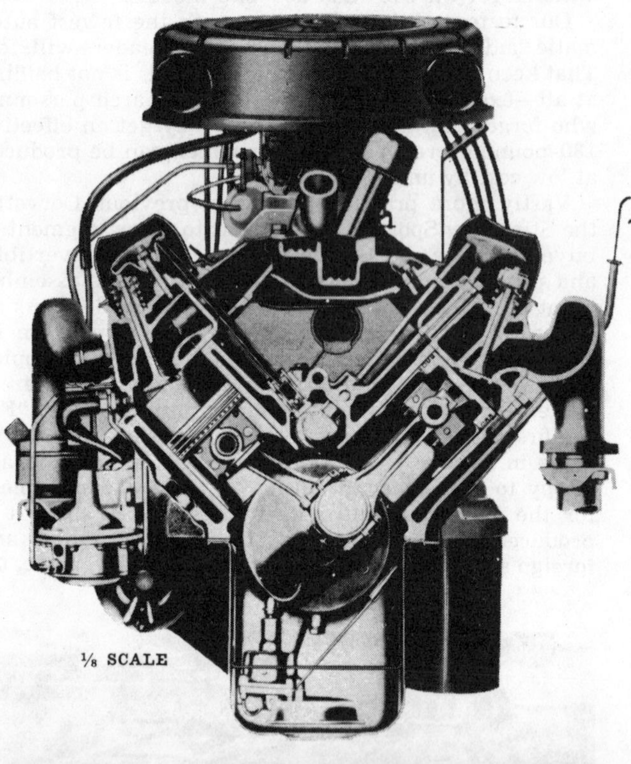

1/8 SCALE

PRICES

Basic price $4,252

OPERATING SCHEDULE

Fuel recommended .... Premium (99-101 Octane)
Mileage .... 10-18 mpg
Range on 20-gallon tank .... 200-360 miles

Oil recommended	Single grade	Multi-grade
32° F and over	SAE 20 or 20W	SAE 10W-30
0° F	SAE 10W	SAE 10W-30
below 0° F	SAE 5W	SAE 5W-20

Crankcase capacity .... 5 quarts
Change at intervals of .... 6,000 miles
Number of grease fittings .... 10 (9 with manual steering)
Most frequent maintenance .... Lubrication at every 6,000 miles

ENGINE:

Displacement .... 327 cu in, 5,370 cc
Dimensions .... 8 cyl, 4.00-in bore, 3.25-in stroke
Valve gear: .... Pushrod-operated overhead valves (hydraulic lifters)
Compression ratio .... 10.5 to one
Power (SAE) .... 300 bhp @ 5,000 rpm
Torque .... 360 lb-ft @ 3,200 rpm
Usable range of engine speeds .... 600-5,500 rpm
Carburetion .... Single four-throat Carter WCFB carburetor

CHASSIS:

Wheelbase .... 98 in
Track .... F 56.3 in, R 57.0 in
Length .... 175.3 in
Ground clearance .... 7.5 in
Suspension: F: Ind., coil springs and wishbones, anti-roll bar
R: Ind., lower wishbones and unsplined half-shafts acting as locating members, radius arms and transverse leaf spring
Steering .... Saginaw recirculating ball with power assistance
Turns, lock to lock 3
Turning circle diameter between curbs .... 36 ft
Tire size .... 6.70 x 15
Pressures recommended .... F 24, R 24 psi
Brakes... Delco-Moraine 11-in drums front and rear, 328 sq in swept area
Curb weight (full tank) .... 3,180 lbs
Percentage on the driving wheels 53

DRIVE TRAIN:

Clutch .... Borg & Beck 10-in single dry plate

Gear	Synchro	Ratio	Step	Over-all	1,000 rpm
Rev	No	2.61	—	8.78	−9.0
1st	Yes	2.54	34%	8.52	9.3
2nd	Yes	1.89	25%	6.36	12.4
3rd	Yes	1.51	51%	5.08	15.6
4th	Yes	1.00	—	3.36	23.5

Final drive ratio .... 3.36 to one

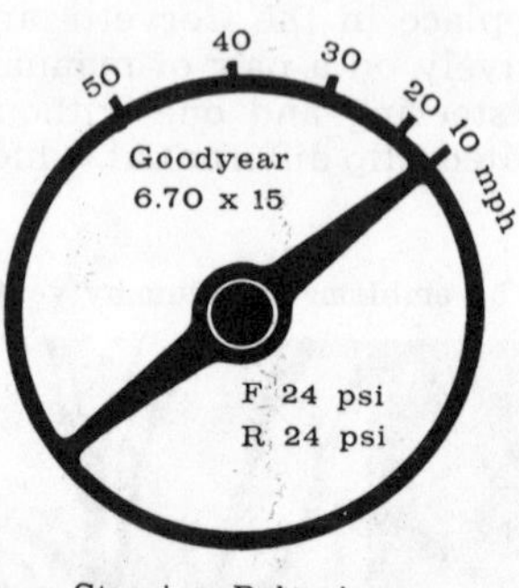

Steering Behavior
Wheel position to maintain 400-foot circle at speeds indicated.

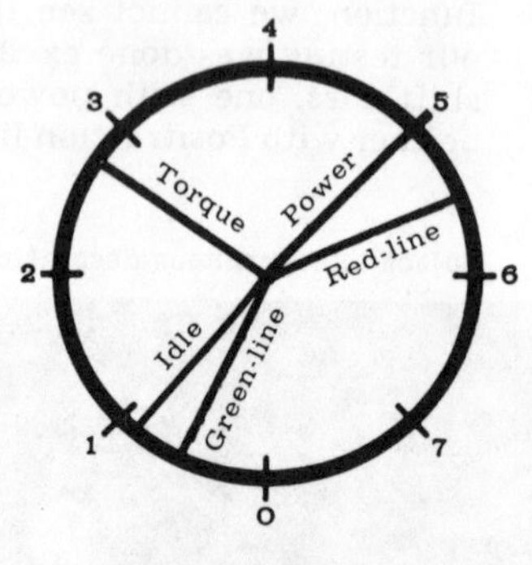

Engine Flexibility
RPM in thousands

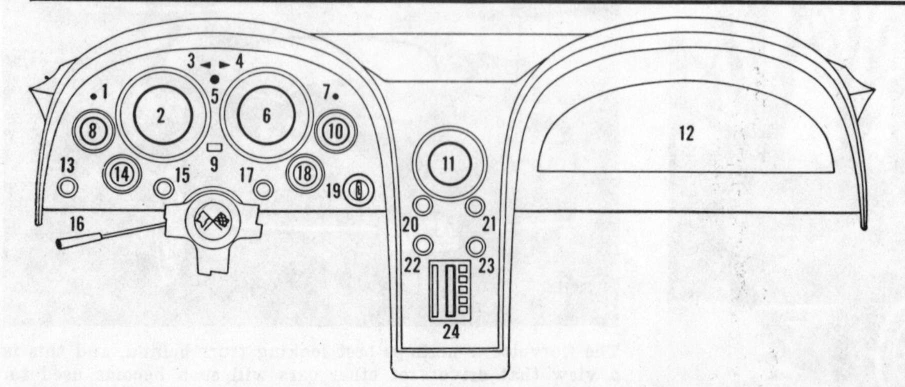

(1) Turn signal warning light (left); (2) Speedometer and odometer; (3) Warning light for headlights on in closed position; (4) Parking brake warning light; (5) High beam warning light; (6) Tachometer; (7) Turn signal warning light (right); (8) Water temperature gauge; (9) Trip odometer; (10) Oil pressure gauge; (11) Clock; (12) Glove box; (13) Light switch; (14) Ammeter; (15) Windshield wiper and washer; (16) Turn signal lever; (17) Cigarette lighter; (18) Fuel gauge; (19) Ignition key and starter; (20) Heater fan and fresh air control; (21) Defroster control; (22) Radio volume and tone control; (23) Radio tuning selector; (24) Radio dial.

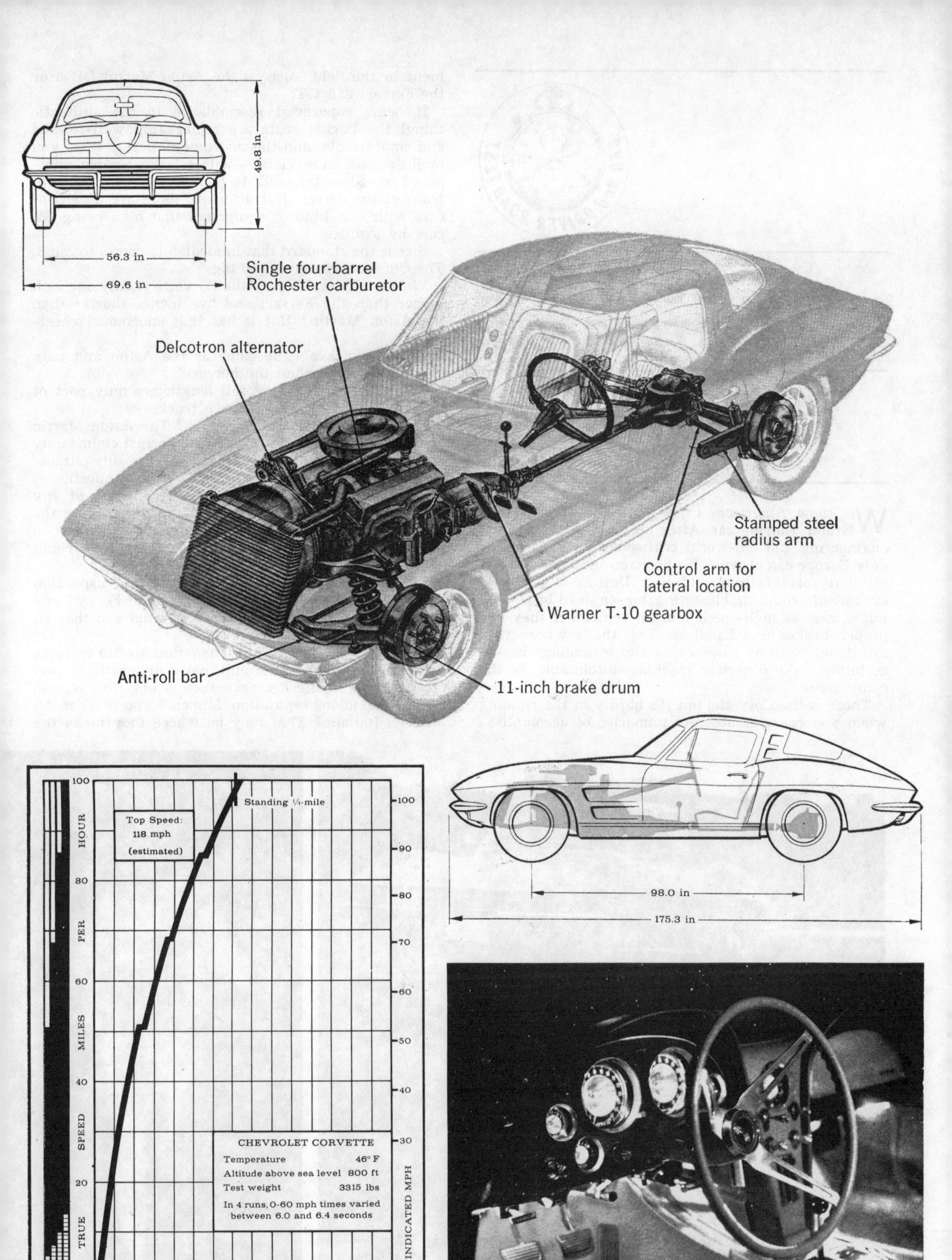

49.8 in
56.3 in
69.6 in
Single four-barrel
Rochester carburetor
Delcotron alternator
Stamped steel
radius arm
Control arm for
lateral location
Warner T-10 gearbox
Anti-roll bar
11-inch brake drum
98.0 in
175.3 in
Standing ¼-mile
Top Speed:
118 mph
(estimated)
CHEVROLET CORVETTE
Temperature 46° F
Altitude above sea level 800 ft
Test weight 3315 lbs
In 4 runs, 0-60 mph times varied
between 6.0 and 6.4 seconds
TRUE SPEED MILES PER HOUR
INDICATED MPH
ACCELERATION TIME SECONDS
1234
SPEED
RANGES
0 2 4 6 8 10 12 14 16 18 20 22 24 26 28 30
0 20 40 60 80 100
30 40 50 60 70 80 90 100

CORVETTE STING RAY

You aren't suggesting that it's one of the best GT cars in the world. Or are you?

We know Americans can't build the best possible Grand Touring car. After all, that takes brilliant engineering and old-world craftsmanship of the kind only Europe can provide. It's heritage and breeding and all that sort of thing, old boy. Detroit is perfectly capable of producing blustering, big-engined buses that might pass as high-speed touring vehicles if they had proper brakes and handling. Yes, that's where they fall down. Squishy suspensions and incendiary brakes certainly make American machines intolerable for the connoisseur.

There is the Corvette, but it's hardly in the running when you consider absolute pinnacles of accomplishment in the field, such as the Aston Martin DB-5 or the Ferrari 250/GT.

It bears superficial resemblance to the authentic thing; the bucket seats are remarkably well-formed and comfortable, and the instrumentation is what you might expect on a GT car—with a large, clearly-numbered speedometer and tachometer placed right in front of the driver. But after all, there are dozens of cars with that kind of equipment that miss being GT cars by a mile.

Size is the standard that immediately comes to mind. The Corvette is simply too big.

Wait a minute. You say the Corvette is only one inch longer than the Ferrari and five inches *shorter* than the Aston Martin? But it has that enormous wheelbase . . .

Only 98 inches? The same as the Aston and only 3.5 inches longer than the Ferrari?

But wheelbase and overall length are only part of it. The Corvette is heavy like a truck.

Its curb weight is only 3180 lbs? The Aston Martin weighs 3450 in the same trim? But Ferrari claims only 2540 lbs for the 250/GT. Bulk and size are silly parameters in such a discussion anyway. The real question is engineering sophistication. That is the heart of any automobile of this sort. After all, the Ferrari and the Aston Martin cost over $12,000, and it's ridiculous to expect that a car less than half as expensive could compete. Take the suspension, for example.

The Corvette has a fully independent suspension system fore and aft? Yes, we know. The Ferrari and the Aston use *live* rear axles? Now what can that all mean?

Then there is the gearbox. The Aston Martin features an optional ZF five-speed unit that is unparalleled, and then, of course, the Ferrari's, though only four-speed, has a tremendous reputation. Muncie? You don't mean Muncie, Indiana? That may be where they make the

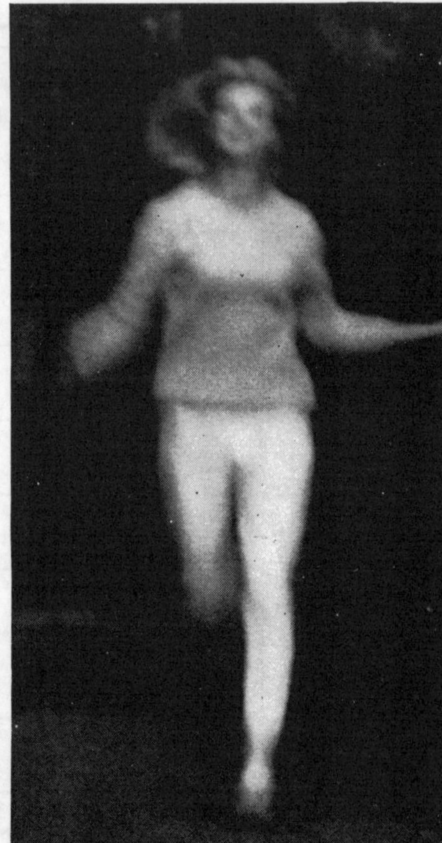

CHEVROLET CORVETTE

Manufacturer: Chevrolet Motor Division.
General Motors Corporation
Detroit 2, Michigan

Price as tested: 5276.30 FOB, St. Louis

ACCELERATION

Zero to	Seconds
30 mph	2.1
40 mph	3.0
50 mph	4.3
60 mph	6.2
70 mph	8.3
80 mph	11.1
90 mph	13.7
100 mph	16.9
Standing ¼-mile	94 mph in 14.9

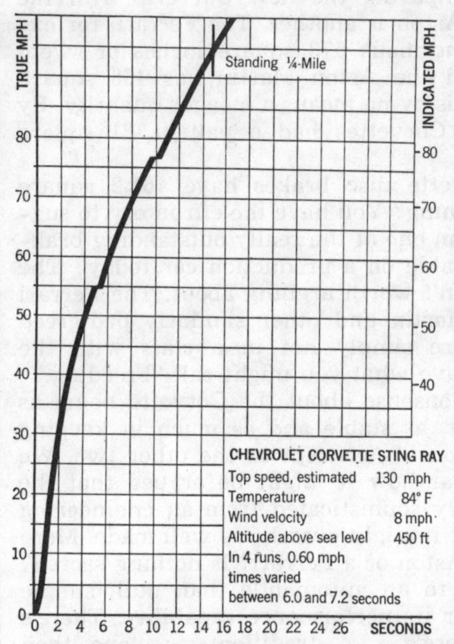

ENGINE

Water-cooled V-8, cast iron block, 5 main bearings
Bore x stroke.....4.00 x 3.25 in, 102 x 83 mm
Displacement.....327 cu in, 5359 cc
Compression ratio.....11.0-to-one
Carburetion.....Single four-throat Holley
Valve gear.....Pushrod-operated overhead valves (hydraulic lifters)
Power (SAE).....350 bhp @ 5500 rpm
Torque.....360 lbs-ft @ 3600 rpm
Specific power output.....1.07 bhp per cu in, 65.3 bhp per liter
Usable range of engine speeds.....600–5600 rpm
Electrical system.....12-volt, 61 amp-hr battery
Fuel recommended.....Premium
Mileage.....10–18 mpg
Range on 20-gallon tank.....200–360 miles

DRIVE TRAIN

Clutch.....10-inch single dry plate
Transmission.....4-speed all-synchro (Muncie)

Gear	Ratio	Over-all	mph/1000 rpm	Max mph
Rev	2.27	7.52	—10.5	—59
1st	2.20	7.40	10.7	60
2nd	1.64	5.51	14.4	77
3rd	1.28	4.30	18.4	102
4th	1.00	3.36	23.5	130

Final drive ratio.....3.36 to one

CHASSIS

Channel-section steel frame, fiberglass body.
Wheelbase.....98 in
Track.....F 56.8 R 57.6 in
Length.....175 in
Width.....69.6 in
Height.....49.8 in
Ground clearance.....7.0 in
Dry weight.....2975 lbs
Curb weight.....3180 lbs
Test weight.....3500 lbs
Weight distribution front/rear.....47/53%
Pounds per bhp (test weight).....10.0
Suspension F Ind., unequal-length wishbones and coil springs, stabilizer bar.
R Ind., radius arms and lower transverse rods, half-shafts acting as upper locating members, transverse leaf spring.
Brakes.....11.75-in discs front and rear, 461.2 sq in swept area
Steering.....Recirculating ball
Turns, lock to lock.....3
Turning circle.....36 ft
Tires.....7.75–15
Revs per mile.....760

CHECK LIST

ENGINE
Starting.....Good
Response.....Excellent
Noise.....Good
Vibration.....Good

DRIVE TRAIN
Clutch action.....Good
Transmission linkage.....Excellent
Synchromesh action.....Excellent
Power-to-ground transmission.....Excellent

BRAKES
Response.....Excellent
Pedal pressure.....Excellent
Fade resistance.....Excellent
Smoothness.....Excellent
Directional stability.....Excellent

STEERING
Response.....Good
Accuracy.....Good
Feedback.....Good
Road feel.....Good

SUSPENSION
Harshness control.....Good
Roll stiffness.....Good
Tracking.....Good
Pitch control.....Good
Shock damping.....Good

CONTROLS
Location.....Good
Relationship.....Good
Small controls.....Good

INTERIOR
Visibility.....Fair
Instrumentation.....Excellent
Lighting.....Good
Entry/exit.....Fair
Front seating comfort.....Good
Front seating room.....Good
Rear seating comfort.....—
Rear seating room.....—
Storage space.....Fair
Wind noise.....Fair
Road noise.....Good

WEATHER PROTECTION
Heater.....Excellent
Defroster.....Excellent
Ventilation.....Good
Weather sealing.....Good
Windshield wiper action.....Good

QUALITY CONTROL
Materials, exterior.....Good
Materials, interior.....Good
Exterior finish.....Good
Interior finish.....Good
Hardware and trim.....Good

GENERAL
Service accessibility.....Fair
Luggage space.....Fair
Bumper protection.....Good
Exterior lighting.....Good
Resistance to crosswinds.....Good

new General Motors four-speed, but who ever heard of a true Grand Touring car with a gearbox made in Muncie, Indiana? It's utterly barbaric. Maybe it does have one of the lightest, most positive linkages ever designed and near-perfect ratios, but it's nonetheless downright silly for *any* transmission to be built in Muncie, Indiana.

When it comes to powerplants, the question is academic. Corvette engines simply can't compete with the modern, overhead camshaft units of the Ferrari and the Aston Martin. Yes, yes, we know all about the Ferrari V-12 dating back to the 1940s, but it *is* one of the greatest designs of all time. And the Aston Martin engine is a lovely double-overhead-cam straight-six. Of course we know the big 365-hp and 375-hp Corvettes have a considerable edge in sheer power, but they're harsh, solid-lifter, semi-racing engines that hardly fit the mold of a smooth, silent GT powerplant.

A new engine for the Corvette? With hydraulic lifters? Three hundred-and-fifty horsepower at 5800 rpm? Silky-smooth? No rough idle? No pushrod clatter? One hundred more horsepower than the 250/GT? Sixty-eight more than the Aston-Martin? That all may be true, but consider the vast advantage the Corvette has in cubic inches. Let's talk in terms of engine efficiency. The Ferrari produces 1.36 horsepower per cubic inch, the Aston Martin gives 1.16 hp per cubic inch and the Corvette 1.072 horsepower per cubic inch. See what we told you? And let's not hear any nonsense about reliability and lack of temperament. When we're discussing *pure* machinery, mundane things of that nature have no bearing whatsoever. Speed in excess of 125 mph is essential for a car of this sort and on that count we'll have to give the Corvette a passing grade. With a 3.31 rear axle ratio, it will easily exceed that mark and, equipped with the optional 3.01 ratio, we grudgingly admit that the machine will top 150 mph without effort.

Anything that goes that fast must certainly be able to stop properly and we all know about those Detroit brakes, don't we?

Disc brakes? On the Corvette? On all four wheels?

As a matter of fact, we do recall reading some press release about Corvette discs, but surely they can't compare . . .

Vented discs, 11¾ inches in diameter? Great resistance to fade? An absolute revelation when compared to the old Corvette brakes?

That may be, but comparing old and new Corvettes is one thing; comparing the new Corvette with the Ferrari and the Aston is another. The Ferrari, for example, has an enormous 573 square inches of swept braking area and the Aston Martin has 468 square inches . . . which is by no means a meager quantity. By contrast the old Corvettes had a paltry 321 square inches.

The new Corvette disc brakes have 461.2 square inches of swept lining? You have the effrontery to suggest that they form one of the really outstanding braking systems available on a production car today? The entire question isn't worth arguing about. The Ferrari and the Aston Martin and other similarly pedigreed European cars are simply not in a class with the Corvette. They have what you might call "breeding."

Stop all that nonsense about the Corvette being as fast and as silent, as stable and as much in keeping with the grand touring concept as the other two. We don't want to hear how it might be argued that the Corvette is equally sophisticated from an engineering standpoint or that it might even be as well made. More reliable than an Aston or a Ferrari? Is nothing sacred?

There is more to an automobile than dull, simple economic value or its performance capability. There's tradition and there's . . . tradition . . . and then there's . . .

Anyway, you know what we're trying to say. **C/D**

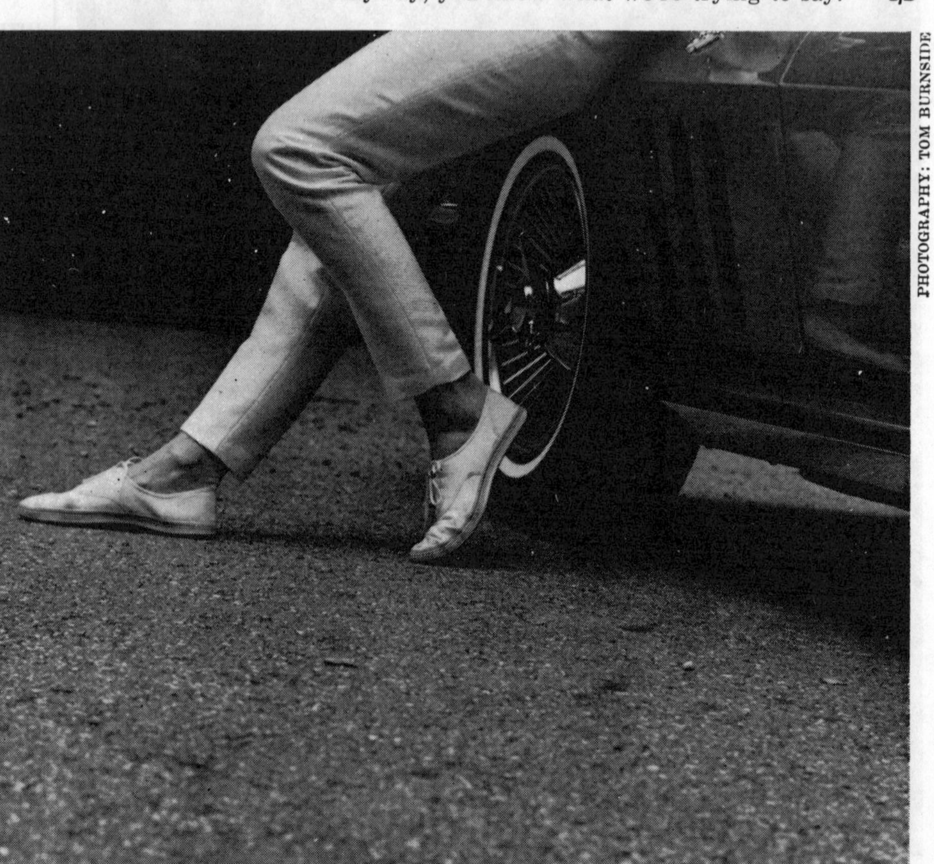

PHOTOGRAPHY: TOM BURNSIDE

CORVETTE STING RAY 427

Next best thing to a psychiatrist—with electric windows, AM-FM, and about a jillion bhp

Son of a Gun—just what the Corvette needed, more power! Chevrolet has replaced last year's top-of-the-line 396-cu. in. engine with a 427. According to the factory, horsepower remains unchanged at 425, but torque has gone up from 415 lbs/ft. at 4000, to 465 at the same rpm. We asked one Chevrolet man why the increase in displacement and torque didn't have a parallel increase in horsepower. "This was done primarily to save weight," he said, with a twinkling eye; "you must remember that cast-iron is very heavy, and by removing thirty cubic inches of it, we have made a significant reduction in weight."

Last year's 396 "porcupine head" Corvette was cranking out quite a bit more than its advertised 425 bhp, and with 427 cu. in., the gap between advertised and actual becomes even broader. However, Chevrolet insists that there are only 425 horses in there, and we'll just have to take their word for it. Though we feel compelled to point out that these are 425 horses of a size and strength never before seen by man—horses as tall as houses, with hooves as big as bushel baskets. When you have *this* many of *those* horses exerting their full force against the small of your back, you are profoundly impressed, and you will most likely lose all interest in counting anyway.

Last month we said that the most dominant feature

of any Ferrari was its engine. Well the same thing is true of this big 427 Sting Ray, except that we'd go one step farther and say that it's the *power,* more than the engine, that overwhelms every other sensation. There's power literally everywhere, great gobs of steam-locomotive, earth-moving torque. And because the Corvette is a relatively small object, and considerably lighter than the other American production cars that are fitted with such engines, there's a direct, one-to-one relationship between the amount of throttle opening and the physical and emotional sensations the driver feels.

Other Corvettes, either the current models with 327 cu. in. engines, or the older ones with the original 283, had a more "European feel." They were very powerful, but they had a zippy, high-winding, no-flywheel feel to them that's missing from the big 427. According to Zora Arkus-Duntov, Chevrolet's Corvette specialist from the very beginning, this difference in sensation is directly related to the comparative torque curves of the two engines. The hottest version of the 327, the 375-bhp fuel-injection engine, produced 350 lbs/ft of torque at 4600 rpm. The new 427 produces 465 lbs/ft at 4000 rpm. As you can see, the 327's *rate* of acceleration was still increasing after the 427's torque curve had dropped off. Thus, even though the 427 is delivering more torque and more acceleration, its earlier torque peak results in less *sensation* of acceleration in the higher rpm ranges.

Except for compensatory suspension changes and a few other twists and improvements, the 427 Sting Ray is basically identical to any other Sting Ray. But the big jump in power that came with the jump from 327 to 427 makes those detail changes critical.

Higher-rate front and rear springs are fitted, along with a ⅞-inch stabilizer bar at the front and a ¾-inch bar at the rear. The half-shafts and U-joints are shot-peened, and are constructed of stronger stuff than on the 327, and there's been in increase in both coolant and lubricant capacity. The 427 can only be ordered with the "Muncie" four-speed transmission and Chevrolet's "Positraction" limited-slip differential. All prudent precautions, which underline the massive thrust available from this new powerplant.

In addition to our 425-bhp test car, Chevrolet also offers a 390-bhp version of the 427-cu. in. engine. This engine has different heads with smaller intake ports, and a smaller-section intake manifold which is cast-iron, as opposed to aluminum on the 425. It also has hydraulic lifters instead of the mechanical lifters in the hotter engine. Both have single four-barrel Holley carburetors, and there's no difference in exterior appearance, save for various emblems and decals and things. The hydraulic-lifter engine develops 390 bhp at 5400 rpm, and 465 lbs/ft of torque at 3600. It's not what you'd call puny, by any stretch of the imagination. In fact, without a watch or a measured quarter-mile, the average driver would have a hard time telling the difference—except for a smoother idle and less mechanical noise.

The interior appearance has no significant changes either. It is still roomy, comfortable, and very well sealed against wind and weather. The seats have a broad range of adjustment, and though they're not the super-buckets of a Ferrari GTB, they are pleasant to sit in for long drives. The back rests still are not adjustable, and this is too bad, especially now that virtually every medium-to-high-priced sedan in the GM line-up offers that feature.

The driving controls and small switches are just fine. The steering wheel telescopes in and out, and this, combined with the seat adjustment, makes it possible to have a pretty decent driving position. Our test car had power brakes and power steering, and we were grateful for both. GM's Saginaw Steering Gear Division has worked hard to provide a normal American-style power steering gear that has some feel and accuracy, and they've done a good job. The Mercedes-Benz system was their performance-target and though the Corvette doesn't measure up quite to that standard (you're still turning a valve, instead of the wheels), it *is* superior to any other American car. The power brakes are even better. They're smooth, free from any grabbiness or directional instability, and they do get it stopped! Even bringing it down from speeds in the 130-140 bracket without any sweat.

Driving the car on an unrestricted proving ground road is a memorable experience. It accelerates from zero to 100 in less than eleven seconds—faster than a lot of very acceptable cars can get to sixty—and is so smooth and controllable in the three-figure speed ranges that it all becomes sort of unreal. In fact, in those circumstances it's pretty hard to tell anything about the car at all, except that it goes like bloody hell and stops and steers without scaring you.

Everything comes into focus when you get it out on the public roads. Compared to anything you might come up against—unless you're unlucky enough to encounter a Cobra 427 (see page 37)—it's the wildest, hottest set-up going. With the normal 3.36 rear axle ratio it'll turn a quarter mile that'll give a GTO morning sickness, and *still* run a top speed of around 150 mph. The 327-engined version is still our favorite, but if you must go faster than anybody else, and you insist upon being comfortable, this is a pretty wild way to go.

The difference between this seven-liter street machine and all the big seven-liter super stocks is in size and proper suspension. All that murderous acceleration is balanced by excellent, almost-lightweight handling. It's stiff and stable and it gets the power on the road—when the wheels stop spinning. First gear is apt to be all wheelspin if you're not careful, and second is almost as bad. Even third can break the rear end loose, if you're down below the 4000-rpm torque-peak, and there's hardly a road in America where you won't be—since that's the equivalent of something in excess of 70 mph! Judicious applications of throttle will eliminate most of the spinning, but there just isn't any way to avoid it in first.

The extra weight of this big engine doesn't really seem to affect the car's handling at all. There's a general feeling of ponderousness that one associates with any of the bigger sports machines at low speeds, but when you're going fast it's quick and responsive. It *is* more difficult to accurately place it in a fast corner, but this is more due to its power steering than to its bulk.

There's no sense even trying to make MGB or Porsche comparisons with this Corvette, because . . . well, because it's so uniquely *American.* It weighs over 3300 pounds but it'll do 140 before most European sports machines are out of second. It does not mince tidily around corners, but it gets around corners faster than most of its peers. It's an *American* GT car, and it'll hold its own in any company at any price.

The most fascinating thing about the continuing success of the Corvette, and the powerful appeal of this new seven-liter contrivance, is its relationship to the widely-publicized GM ban on racing. Many people thought that the Corvette would wither and die when it was no longer the car to beat in the races. To the contrary, they're selling better than ever—and going better than ever as well. It's a shame really, if they keep building them this good and this fast, they may *never* have to go racing again. **C/D**

PHOTOGRAPHY: SUDNIK & OVERHARDT

SPECIFICATIONS OVERLEAF

CORVETTE STING RAY 427

Manufacturer: Chevrolet Motor Division
General Motors Corporation
Detroit 2, Michigan

Price as tested: ca. $6000

ACCELERATION

Zero to	Seconds
30 mph	3.2
40 mph	3.8
50 mph	4.6
60 mph	5.4
70 mph	6.9
80 mph	7.8
90 mph	9.0
100 mph	10.6
Standing ¼-mile	112 mph in 12.8

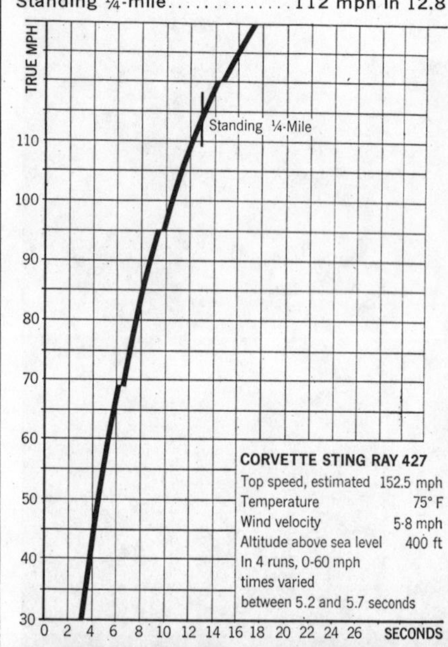

ENGINE

Water-cooled V-8, cast iron block, 5 main bearings

Bore x stroke.....4.25 x 3.76 in, 106 x 94 mm

Displacement...............427 cu in, 7000 cc

Compression ratio..................11.0 to one

Carburetion.................Single 4-bbl Holley

Valve gear.........Pushrod-operated overhead valves, mechanical lifters

Power (SAE).............425 bhp @ 5600 rpm

Torque.................460 lbs/ft @ 4000 rpm

Specific power output.....0.995 bhp per cu in, 60.7 bhp per liter

Usable range of engine speeds.500–6500 rpm

Electrical system...12-volt, 61 amp-hr battery, 37A alternator

Fuel recommended..................Premium

Mileage..............................10–15 mpg

Range on 20-gallon tank........200–300 miles

DRIVE TRAIN

Clutch...............10.5-inch single dry plate

Transmission.............4-speed, all synchro

Gear	Ratio	Overall	mph/1000 rpm	Max mph
Rev	2.26	7.59	−10.3	−66
1st	2.20	7.39	10.7	69
2nd	1.64	5.51	14.3	95
3rd	1.27	4.27	18.5	120
4th	1.00	3.36	23.5	152.5

Final drive ratio.....................3.36 to one

CHASSIS

Wheelbase...............................98 in

Track.........................F 56.8, R 57.6 in

Length....................................175 in

Width.....................................69.6 in

Height....................................49.8 in

Ground Clearance......................7.0 in

Dry weight..............................3005 lbs

Curb weight.............................3160 lbs

Test weight..............................3380 lbs

Weight distribution front/rear..........47/53%

Pounds per bhp (test weight)...............7.9

Suspension F: Ind., unequal length wishbones, coil springs, anti-sway bar.
R: Ind., lower transverse link, half-shafts acting as upper locating members, trailing arms, transverse leaf spring.

Brakes..........11.75-in discs, front and rear, 461.2 sq in swept area

Steering.......................Recirculating ball

Turns, lock to lock...............................3

Turning circle...............................36 ft

Tires...................................7.75 x 15

Revs per mile...............................760

CHECK LIST

ENGINE

Starting.............................Good

Response...........................Good

Noise.................................Fair

Vibration............................Fair

DRIVE TRAIN

Clutch action..................Excellent

Transmission linkage..........Excellent

Synchromesh action...........Excellent

Power-to-ground transmission.......Fair

BRAKES

Response......................Excellent

Pedal pressure.................Excellent

Fade resistance................Excellent

Smoothness....................Excellent

Directional stability............Excellent

STEERING

Response..........................Good

Accuracy...........................Fair

Feedback...........................Fair

Road feel..........................Good

SUSPENSION

Harshness control.................Good

Roll stiffness......................Good

Tracking............................Good

Pitch control.......................Good

Shock damping.....................Good

CONTROLS

Location...........................Good

Relationship.......................Good

Small controls.....................Good

INTERIOR

Visibility...........................Good

Instrumentation...............Excellent

Lighting............................Good

Entry/exit...........................Fair

Front seating comfort.............Good

Front seating room................Good

Rear seating comfort.................—

Rear seating room....................—

Storage space......................Poor

Wind noise..........................Fair

Road noise..........................Fair

WEATHER PROTECTION

Heater..........................Excellent

Defroster.......................Excellent

Ventilation.........................Good

Weather sealing...................Good

Windshield wiper action...........Good

QUALITY CONTROL

Materials, exterior................Good

Materials, interior.................Good

Exterior finish.....................Good

Interior finish......................Good

Hardware and trim.................Good

GENERAL

Service accessibility...............Good

Luggage space.....................Poor

Bumper protection.................Good

Exterior lighting...................Good

Resistance to crosswinds..........Good

Some purists snickered in 1953 when Chevrolet brought out a fiber glass sports car with an automatic transmission. They've been strangely silent lately.

Maybe that's because Corvette has a whole lot to say for itself. It comes with the kind of design features you know how to use to demonstrate traffic safety. Things like IRS so surefooted and sophisticated it makes the suspension on some of those $14,000 imports seem behind the times. Big disc brakes all around. Fully synchronized four-speed box if you'd rather stir your gears by hand. (A fully synchronized three-speed is standard with the 300 horsepower engine.) V8s you can order with up to 427 cubic inches and 425 horsepower. And lines that mate function to form. Maybe that's the reason for all the silence. Or maybe a little bird told those purists something about fiber glass sports cars with automatic transmissions.

Sting Ray Coupe.

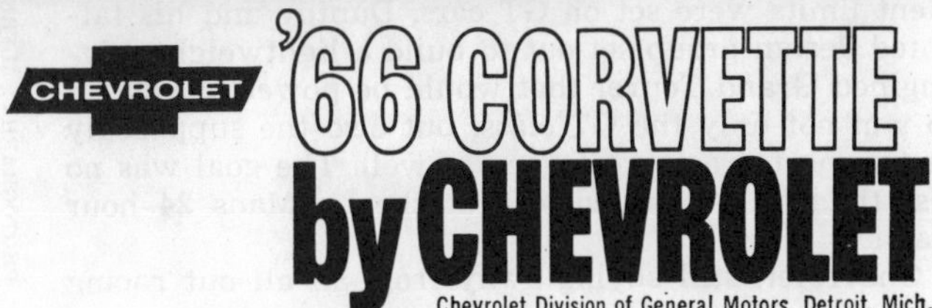

Chevrolet Division of General Motors, Detroit, Mich.

David Erwin of Painted Post, New York, describes himself as a "Chevy man." There are "Ford men" and "Plymouth men" and "Pontiac men," but Erwin and the close-knit group of friends who gather nightly to labor over one of the rarest of all Chevrolets remain loyal to that marque.

David Erwin could also be cast as a "Ferrari man." A tall, reserved banker, and the heir of a prominent family of landed gentry whose holdings date back to the Revolution, Erwin is the sort of precise, well-bred young man whose fascination with the Maranello product is practically a foregone conclusion. In fact, one stall in Erwin's comfortable workshop behind the family's large, colonnaded homestead is occupied by one of the last GTO lightweights ever built.

But the machine stabled beside the Ferrari is the one that keeps "Chevy man" Erwin and his friends turned on. This is one of the fabled "Grand Sport" race cars, a four-wheeled passenger pidgeon whose existence bears testimony to those mystery-shrouded days when Chevrolet was committed to winning every major sports car race in the world, including Le Mans.

Ever since the spring of 1963, when General Motors summarily cancelled all competition activities, the Corvette Grand Sport has established itself as one of the most fascinating enigmas in motor racing annals. Periodically, one of the five examples Chevy built has appeared at places like Sebring or Nassau, has raced around at shocking speeds, and then has plunged back into mysterious seclusion. People are still talking about how the Jim Hall/Roger Penske Grand Sport stunned the Fords and Ferraris by leading the early laps of the 1964 Sebring 12-Hour, and how Penske thrashed the late Ken Miles and his Cobra 427 prototype at Nassau later the same year.

These rare outings, though impressive, never did the machines full justice, because GM's anti-racing policy was so effective than never once did a Grand Sport reach the race track at the level of readiness that Chevrolet engineers had intended. That the cars were so impressive when they were finally raced—both outdated and underpowered—can only make one pause to wonder how overwhelming they might have been if Chevrolet had been permitted to carry out the full Grand Sport program with corporate blessing.

The Grand Sport was a direct development of the 1958 FIA ruling that limited international sports/racing cars to three liters engine displacement. Up until that moment, Chevrolet had been hard at work on some exotic big-engined sports/racers, the last of which, the Corvette SS, ran at the 1957 Sebring race in the hands of Piero Taruffi and John Fitch. Seeing no benefit to passenger-car engineering in perfecting a 3-liter racing engine, Chevrolet stayed away from road racing until their resident competition wizard, Zora Arkus-Duntov, spotted a loophole in the FIA rules that would permit Chevrolet's return. Because no displacement limits were set on GT cars, Duntov and his talented design group set out to build a lightweight, big-engined Grand Tourer that would be powerful enough to win not only the GT class, but also the supposedly faster sports/racing category as well. The goal was no less than an overall victory at the Le Mans 24-hour classic.

Chevrolet, still shying away from an all-out racing

GRAND SPORT

BY BROCK YATES

Long ago in th
General Motor

PHOTOGRAPHY: ALFRED FISHER

e mysterious days when
vas racing, they built a car...

car disguised as a GT car, wanted their Grand Sport to look like a production car—in this case, their brand new Sting Ray. Duntov and his Corvette engineers reckoned that it would take 600 horsepower to push the Sting Ray up the straight at Le Mans at a competitive top speed, 4-wheel disc brakes to slow it down, and a vehicle weight of 1800 lbs. to achieve competitive lap times.

Because the rules allowed a GT engine to displace less than its stated capacity, but not more, Chevrolet specified a 327 cu. in. aluminum block with the standard 4-inch bore and a ¾-inch stroker crank (4″ x 4″, in hot rod parlance), for an enormous 402 cu. in. displacement. The heads, also aluminum, have never been seen by the public. The combustion chambers were hemispherical and featured twin ignition. Sitting astride the engine was a complex fuel injection system, with eight long ram tubes poking up through the hood.

Initial tests proved the 4″ x 4″ setup unsatisfactory and subsequent engines were given 3.75-in. strokes, which reduced displacement to 377 cu. in.

The chassis was tubular aluminum with the fully independent Sting Ray rear suspension. The body was slightly smaller than stock Sting Ray to improve the

CORVETTE GRAND SPORT SPECIFICATIONS*

Manufacturer: Chevrolet Motor Division
General Motors Corporation
Detroit, Michigan

Vehicle type: Front-engine, rear wheel drive, 2-passenger GT racing car, ladder frame, aluminum structural members integrated with reinforced plastic body, plexiglass side and rear windows

ENGINE:

Type: Water-cooled V-8, aluminum block and heads, 5 main bearings, twin ignition, hemispherical combustion chambers, two valves per cylinder
Bore x stroke....................4.00 x 3.75 inches, 101.7 x 95.3 mm
Displacement....................377 cu inches, 6172 cc
Compression ratio....11.0 to one (alternative ratios: 10.0, 12.0 to one)
Carburetion....................Port-type fuel injection
Valve gear........Push-rod operated overhead valves, mechanical lifters
Valve timing: Intake opens....................55° BTDC
Intake closes....................117° ABDC
Exhaust opens....................106° BBDC
Exhaust closes....................44° ATDC
Valve lift....................0.46 inches (intake and exhaust)
Valve diameter: Intake 2.20 inches
Exhaust 1.72 inches
Power (SAE)....................550 bhp @ 6400 rpm**
Torque (SAE)....................500 lbs/ft @ 5200 rpm**
Specific power output....................1.46 bhp/cu in, 89.3 bhp/liter
Maximum recommended engine speed....................6500 rpm**

DRIVE TRAIN

Transmission....................4-speed manual, all synchromesh
Clutch diameter....................10.0 inches
Final drive ratio... 3.55 (alternative ratios: 3.08, 3.36, 3.70, 4.11, 4.56)

Gear	Ratio	mph/1000 rpm	Max. speed (3.55 ratio axle)
I	2.21	10.6	69 mph @ 6500 rpm
II	1.64	14.2	92 mph @ 6500 rpm
III	1.27	18.4	119 mph @ 6500 rpm
IV	1.00	23.3	152 mph @ 6500 rpm

DIMENSIONS AND CAPACITIES

Wheelbase....................98.0 in
Track....................F: 56.8 in, R: 57.8 in
Length....................172.8 in
Width....................69.6 in
Height....................51.9 in
Ground clearance....................4.3 in
Curb weight....................1908 lbs
Lbs/hp....................2.9
Battery capacity....................12 volts, 61 amp/hr
Generator capacity....................624 watts
Fuel capacity....................36 gal
Oil capacity....................9 qts
Water capacity....................16.5 qts

SUSPENSION

F: Ind., fabricated tubular unequal-length wishbones, combined coil spring/shock absorber units, anti-sway bar
R: Ind., lower transverse leaf spring, half-shafts acting as upper locating links, trailing arms, tubular hydraulic shock absorbers

STEERING

Type....................Recirculating ball
Turns lock-to-lock....................3.25
Turning circle....................40 ft

BRAKES

F: 11.47-in Girling discs with 3.30 x 2.04-in pads (power assisted)
R: 11.50-in Girling discs with 3.06 x 1.55-in pads (power assisted)
Swept area....................435.2 sq inches

WHEELS AND TIRES

Wheel size and type..6.0 x 15-in, magnesium knock-off type, weighing 16.5 lbs each
Tire make, size and type........Firestone SS170 TW, F: 7.10/7.60-15
R: up to 8.00/8.20-15

*From FIA Application 837 A-63
**Our estimate

The Grand Sport's engine was to be a 600-hp, 402 cu. in., all-aluminum V-8 with twin ignition and hemi heads.

aerodynamics, although extra-wide fender valances later had to be fitted to accommodate ever-wider tires. Aside from the front headlights being placed behind streamlined plexiglass housings, the cars looked quite similar to the stock Sting Rays.

Chevrolet submitted papers for FIA recognition of the Grand Sport early in 1963, promising to have built 100 examples between July 7, 1962 and June 1, 1963 (Le Mans was on June 15-16 in 1963), but the papers were hastily withdrawn when the Corporation pulled the plug on the racing program. An unauthorized copy of the papers, hoarded for years, turned up in New York recently, and makes interesting reading indeed (see the specifications table, opposite).

As far as we can determine, five coupes were built by Chevy before the racing ban. Chevrolet originally intended to market several hundred production Grand Sports to the public at a price under $10,000, but that aspect of the project never got off the ground, and the "Grand Sport" label was applied only to the factory-built race cars.

Duntov and his crew went to Sebring, Florida, in mid-January '63, for pre-race practice, with Masten Gregory, Dr. Dick Thompson, and Duntov himself do-

(continued from preceding page)

ing the driving. The 377 cu. in. engines were not ready and fuel-injected 327s were used in their stead. It was discovered that the Girling solid discs would last only a few hundred miles, and Chevrolet switched to vented discs of their own design with Girling calipers (presaging the vented discs of the '65 Sting Ray). The test program was judged an overall success, and the Chevrolet people planned to debut the cars at Sebring in March, 1963, followed by an all-out assault on Le Mans that June.

While details for the trip to the Sarthe circuit were being worked out, Mickey Thompson built some lightweight Sting Rays of his own, powered by the top-secret "Mk. II" 427 cu. in. porcupine-head stock car engine, and entered them in the Daytona Continental. In later years, some of the Thompson lightweights were to become confused with the factory-built Grand Sports, but they were quite unrelated to the exquisite machines that Duntov's group was completing in the winter of 1963.

Immediately following the Daytona 500, General Motors lowered the boom. Three Mk. II-engined Chevy stockers driven by Junior Johnson, Johnny Rutherford and G. C. Spencer were so much faster than everything else that only incomplete "debugging" prevented them from turning the race into a private donnybrook. Success seemed a heartbeat away. When GM brass-hats Donner and Gordon suddenly announced that the Corporation would henceforth follow the 1957 Automobile Manufacturers Association anti-racing resolution to the letter, the enthusiasts at Chevrolet were probably as shocked as anybody else. The Grand Sport—and with it Sebring and Le Mans—became as dead an issue to GM as the Missouri Compromise.

The cars sat around in some dark corner of Chevrolet Engineering for 10 months. Then the lid opened a crack before slamming shut for another year. In December of '63, three of the Grand Sports showed up at Nassau. Two were entered by John Mecom, while the other was in Jim Hall's stable. The cars were not seen again until Nassau '64, when Penske scored the first of three wins that week, immediately prior to his retirement.

By now the Grand Sports were getting old. Chevy seemed to have forgotten about them. No eyebrows were raised when two of them showed up at Sebring in March '65. No eyebrows, that is, except Dave Erwin's.

From the moment he clapped eyes on a Grand Sport, Erwin wanted one. He was spellbound at the sight and sound of the wickedly powerful machine. He followed the course of the car at Sebring and Elkhart Lake, after which it came into the hands of Pennsylvanian George Wintersteen. Wintersteen, a friend of Penske's, raced the car sporadically —at Sebring in '66, and in some club events. Last fall Wintersteen casually offered it for sale minus engine, and Erwin snapped it up.

The car was in excellent condition. The plastic headlight covers had been replaced with wire mesh; the differential oil cooler, which had originally been attached to the rear deck lid, was missing; the filler cap had been altered slightly; and the air jacks had been removed. Otherwise the car was practically the same as the day it first rolled out of the General Motors Tech Center nearly five years before.

When we were given the opportunity to drive Erwin's Grand Sport (designated Chassis #5) the car was fitted with a muscled-up 327 that Erwin, his friends, and his brother Tom, had stuffed aboard. The engine—which utilizes a Crower cam and hi-rev kit, General Motors exhaust headers and hi-rise aluminum manifold, cast iron heads from the fuel-injection Corvette engine, and a big Holley carburetor—is intended as an interim powerplant. Last May, Dave saw another ad.

This one was placed by NASCAR short-track star Bobby Allison, and was offering one complete Mk. II engine, a pair of spare blocks and enough extra Mk. II spares to build another complete engine. Erwin grabbed them, and is hoarding them as if they were H. L. Hunt's oil leases.

Because the blue and white brute wasn't registered (Erwin plans to run the car in a few regional races, then restore it, but never drive it on the street), we towed it a short distance to a long stretch of completed but not yet opened four-lane highway for some test runs. The lads had ingeniously stuffed the mufflers designed for the new Sting Ray external side-mounted exhaust pipes into the enormous Grand Sport tubes, which reduced the noise level to a point where the off-duty personnel in the State Police barracks a few miles down the road wouldn't be unduly disturbed.

It was well and truly a racer. Thumping down the highway on the tremendous Firestone Indy tires, the familiar odors of oil and hot paint wafted into the cockpit, along with the sound of air rushing around the hand-operated plexiglass windows. This mingled with the whine of the fully-locked differential gears and the slick prototype Muncie gearbox.

The gearbox and brakes were nearly perfect. That means stops like the car had just run into a mud bank, while the transmission was as loose—and yet precise—as any we've ever handled. The locked rear end made it an awful chore to negotiate corners under 30 mph, mainly because the inside rear wheel would moan and scuff the pavement, and the rear end sounded as if it was going to explode through its cast aluminum housing, but at high speeds the car was a dream. It had virtually neutral steering characteristics, and we could find nothing in its entire handling range that could be described as treacherous or unstable.

Naturally, "Chevy man" Erwin is considering the installation of one of his Mk. II 427s. Although he admits this would not be an entirely authentic switch, he is correct when he says that the assembly of a hemi-head, twin-ignition, port-injected, all-aluminum Chevy engine is out of the question, and he hopes that this will get him off the hook with the historical purists.

In any case, David Erwin of Painted Post has himself one of the most unique automobiles in the world. The other four Grand Sports have been sold and resold several times, although either Wintersteen or Penske has owned all of them at one time or another. Wintersteen and Mecom currently own Grand Sports that have been converted into roadsters, while Texan Delmo Johnson and Toledo, Ohio, Chevy dealer Jim White own the remaining two coupes. So Erwin has one fifth of the total supply of Grand Sports, making his possession all the more valuable.

There it sits, in the loving care of Erwin and his buddies, looking ready to tangle with any GT car built in the half-decade since the car was made. It surely is one of the fastest relics in the world—a car that remains only a few seconds off the best lap times of the fastest road racing cars in the world, even today.

Given an even break, old Chassis #5 might have brought the Corporation that tossed it out like an illegitimate son America's first victory at Le Mans—and three years ahead of time at that. But that's all over now . . . except for "Chevy man" David Erwin. C/D

CAR and DRIVER ROAD TEST

CHEVROLET CORVETTE 427

You voted it the Best All-Around Car of 1967, and we think you just might have something there.

PHOTOGRAPHY: STUDIO PLACE

The Corvette has come a long way since it was introduced in 1953. In the beginning, the Corvette was a cute little two-seater. It sure enough looked like a sports car, but underneath the radical fiberglass bodywork was a puny 235 cu. in., 150-horsepower "Blue Flame" six and a two-speed Powerglide transmission. Everybody laughed. Even Thunderbird owners knew they had something closer to a sports car.

Slowly, the Corvette got better. In 1955 came a V-8 and a 3-speed, close-ratio, all-synchro gearbox. The following year, Dr. Dick Thompson astounded the purists when his Corvette led the production car race at Torrey Pines—fending off the undefeated Mercedes-Benz 300SLs and Jaguar XK-140MCs until the 'Vette's brakes faded him back to a second-place finish.

The Corvette was rolling in high gear by 1957, with the fuel-injected 283 V-8, a 4-speed gearbox and heavy-duty brakes. Winning production car races was easy, and Chevy unveiled a real, live sports/racing car at Sebring, the Corvette SS. The AMA anti-racing ban removed GM from racing shortly thereafter, and the pace of development slowed again, with the only memorable milepost being the 327 cu. in. V-8 in 1961.

Nineteen sixty-three was a banner year for Corvette. A whole new car, the Sting Ray, debuted—the first all-new Corvette in a decade. It shared with the rear-engined Corvair the distinction of being the only American car with four-wheel independent suspension, and it was well and truly a sports car. The emergence of the Sting Ray was underlined by the appearance of the Grand Sport, chronicled in our April issue, during GM's brief return to racing in 1963.

The Sting Ray is the most sophisticated car made in America, and among the best-engineered sports cars made anywhere in the world.

Nobody was laughing anymore. The Corvette was accepted by purist and hot rodder alike. The automotive magazines claimed they could boost circulation by 30,000 (about the number of Corvettes sold each year) merely by putting a Sting Ray on the cover. Even Ford was irked enough to back Carroll Shelby's Cobra in a last-ditch effort to wrest the world's attention away from the Corvette. The Corvette had arrived and everybody in the blue-eyed world knew it.

Many experimental features are first tried out on the low-volume Corvette, then applied to the regular passenger cars, like aluminum transmission cases and bell housings, 4-speed transmissions, rayon brake lines, and link-type independent rear suspension. Standard engines are modified for use in the Corvette, and the modifications often turn up later in regular production Chevies.

Nothing of earth-shattering importance has happened to the Sting Ray in the past four years. Four-wheel disc brakes—an American first—and the 396 cu. in. "porcupine" head engine appeared in 1965, followed by the 427 cu. in. V-8 in 1966. A year ago, it looked as if the Mako Shark show car might prefigure the '67 Corvette, and there were dark rumors about a Chaparral-like rear-engined Corvette, but nothing materialized. Zora Arkus-Duntov, the man most responsible for the Corvette's success (nominally, he is head of the Corvette's engine and chassis group), and his *cadre* of engineers have concentrated on detail refinements, to the obvious satisfaction of not only the nearly 100,000 Corvette owners, but also to a sampling of *Car and Driver's* nearly 400,000 readers, who voted it the Best All-Around Car of 1967—over very stiff competition.

It was the results of our Readers' Choice poll that suggested to us that it was high time we tested another Corvette, but which one? There are two basic body types (the coupe and the convertible hardtop); two basic engines (the 327 and the 427); three basic transmissions (the 2-speed Powerglide and 3- and 4-speed manuals); and a staggering variety of options that can turn the Corvette into anything from a luxury two-seater to an all-out racer—and everything in between. We last tested a Corvette in our October '65 issue; a very strong 427, all *sturm und drang*. Our own personal preference was the 327 cu. in., 350-hp hydraulic lifter engine (*C/D*, January '65) because it is lighter, better balanced, and more responsive than the 427. It's also a lot more practical for everyday street driving. But when we contacted Duntov, he talked us into

With the aluminum cylinder heads, the 427 engine weighs only 40 lbs. more than the cast iron 327. And the "three by two" carburetion is as smooth as fuel injection.

testing a new version of the 427, with a 3 x 2-barrel carburetion set-up and aluminum heads.

This engine, Duntov claimed, is only 40 lbs. heavier than the cast iron 327—and it is weight that is removed from the front end. The cylinder heads alone save 75 lbs. Sure enough, this 427 Corvette weighs 43 lbs. *less* than the last 327 we tested, and it's distributed 46/54% to the 327's 47/53.

The heart of the new carburetion system is a unique air-operated system of controlling the throttle plates of the end carbs; only the center carb is mechanically connected to the accelerator pedal. An all-mechanical linkage bogs down if the throttle is punched at low speeds. Other manufacturers have tried manifold vacuum to control the secondary carburetors, but it has never been as successfully worked out as on this car. The new system operates the secondaries from the venturi vacuum in the primary carb. It results in an astoundingly tractable engine and uncannily smooth engine response. With a venturi area about the size of a barn door, it's possible to drive off in high gear with very little slipping of the clutch or feathering of the throttle. As soon as it's rolling, say at 500 rpm, you can push the throttle to the floor and the car just picks up with a turbine-like swelling surge of power that never misses a beat all the way up to its top speed of over 140 mph. And you get the same response—instantly—in any gear any time you open the tap. The only drawback of the system is when you back off from full throttle at engine speeds over 5500 rpm (1000 rpm under the redline). The secondaries don't close immediately—it takes about 2/10ths of a second—because the air linkage isn't as fast or as positive as mechanical operation. On the whole, the Corvette's three deuces are as smooth and responsive as fuel injection.

This engine is regular production option, coded L89 (so you know what to ask for), but it isn't quite the end of the line. There is an even more powerful, Instant Im-

(Text continued on page **97***:*
Specifications overleaf

CHEVROLET CORVETTE 427

Manufacturer: Chevrolet Motor Division
General Motors Corporation
Detroit, Mich.

Number of dealers in U.S.: 6600

Vehicle type: Front-engine, rear-wheel-drive, 2-passenger sports/GT car, all-steel ladder frame with fiberglass reinforced plastic body

Price as tested: $5900.15
(Manufacturer's suggested retail price, plus options listed below, Federal excise tax, dealer preparation and delivery charges; does not include state and local taxes, license or freight charges)

Options on test car: High-performance 427 cu. in. engine with 3 x 2-bbl. carburetion ($437.10), aluminum cylinder heads ($368.65), tinted glass ($15.80), electric windows ($57.95), limited-slip differential ($42.15), shoulder harnesses ($26.35), power brakes ($42.15), transistor ignition ($73.75), four-speed manual transmission ($184.35), power steering ($94.80), AM/FM radio ($172.75), 7.75-15 red stripe tires ($31.35)

ENGINE
Type: Water-cooled V-8, cast iron block, aluminum cylinder heads, 5 main bearings
Bore x stroke...4.25 x 3.76 in, 108 x 95.5 mm
Displacement..............427 cu in, 6994 cc
Compression ratio................11.0 to one
Carburetion...................3 x 2-bbl. Holley
Valve gear.........Pushrod-operated overhead valves, mechanical lifters
Power (SAE).............435 bhp @ 5800 rpm
Torque (SAE)..........460 lbs/ft @ 4000 rpm
Specific power output..........1.02 bhp/cu in, 62.2 bhp/liter
Max. recommended engine speed...6500 rpm

DRIVE TRAIN
Transmission.....4-speed manual, all-synchro
Clutch diameter.........................11.0 in
Final drive ratio...................3.55 to one

Gear	Ratio	Mph/1000 rpm	Max. test speed
I	2.20	9.9	64 mph (6500 rpm)
II	1.64	13.3	86 mph (6500 rpm)
III	1.27	17.2	112 mph (6500 rpm)
IV	1.00	21.8	142 mph (6500 rpm)

DIMENSIONS AND CAPACITIES
Wheelbase..............................98.0 in
Track...................F: 57:6 in, R: 58.3 in
Length...............................175.1 in
Width..................................69.6 in
Height.................................49.6 in
Ground clearance........................5.0 in
Curb weight..........................3137 lbs
Test weight..........................3563 lbs
Weight distribution, F/R..............46/54%
Lbs/bhp (test weight)......................8.2
Battery capacity...........12 volts, 61 amp/hr
Alternator capacity...................288 watts
Fuel capacity..........................20.0 gal
Oil capacity..............................6.0 qts
Water capacity..........................23.0 qts

SUSPENSION
F: Ind., unequal-length wishbones, coil springs, 0.875-in anti-sway bar
R: Ind., drive shafts acting as upper links, lower lateral links, single trailing arms, transverse leaf spring

STEERING
Type.........................Recirculating ball
Turns lock-to-lock..........................3.0
Turning circle...............................40 ft

BRAKES
F: 11.75 x 1.25-in Delco-Moraine vented discs
R: 11.75 x 1.25-in Delco-Moraine vented discs with integral drum parking brake
Swept area.......................461.2 sq in

WHEELS AND TIRES
Wheel size and type..............6.0JK x 15-in, stamped steel, 5-bolt
Tire make, size and type....Firestone 7.75-15 Speedway 500, two-ply nylon tubeless
Test inflation pressures....F: 24 psi, R: 24 psi
Tire load rating.....1270 lbs per tire @ 24 psi

PERFORMANCE
Zero to	Seconds
30 mph	2.0
40 mph	2.6
50 mph	3.5
60 mph	4.7
70 mph	6.4
80 mph	8.0
90 mph	10.1
100 mph	12.3

Standing ¼-mile.........13.6 sec @ 105 mph
80-0 mph.......................257 ft (.83 G)
Fuel mileage.......9-13 mpg on premium fuel
Cruising range..................180-260 mi

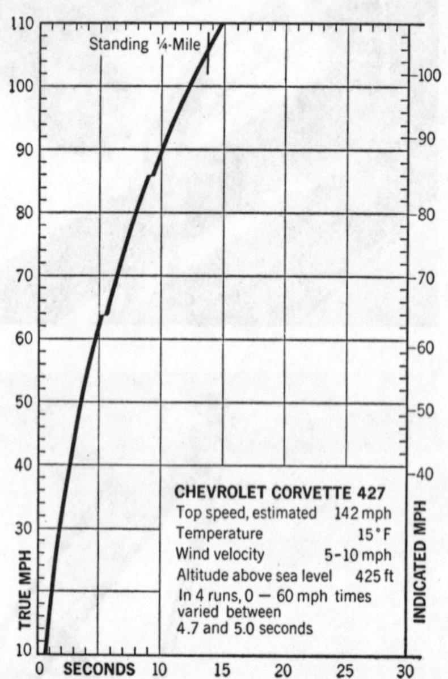

CHECK LIST

ENGINE
Starting..........................Very Good
Response..........................Excellent
Vibration.........................Excellent
Noise.................................Good

DRIVE TRAIN
Shift linkage.....................Excellent
Synchro action....................Excellent
Clutch smoothness.................Excellent
Drive train noise.....................Good

STEERING
Effort............................Very Good
Response..........................Very Good
Road Feel.............................Good
Kickback..........................Very Good

SUSPENSION
Ride Comfort......................Very Good
Roll resistance.......................Good
Pitch control.........................Good
Harshness control.....................Good

HANDLING
Directional control...............Very Good
Predictability....................Very Good
Evasive maneuverability...........Very Good
Resistance to sidewinds...........Very Good

BRAKES
Pedal pressure....................Excellent
Response..........................Excellent
Fade resistance...................Excellent
Directional stability.............Excellent

CONTROLS
Wheel position........................Fair
Pedal position........................Good
Gearshift position....................Good
Relationship..........................Good
Small controls........................Good

INTERIOR
Ease of entry/exit....................Good
Noise level (cruising)................Good
Front seating comfort.................Good
Front leg room........................Fair
Front head room.......................Good
Front hip/shoulder room...............Good
Rear seating comfort.....................—
Rear leg room............................—
Rear head room...........................—
Rear hip/shoulder room...................—
Instrument comprehensiveness......Excellent
Instrument legibility.............Excellent

VISION
Forward...............................Good
Front quarter.........................Good
Side..................................Good
Rear quarter..........................Fair
Rear..................................Fair

WEATHER PROTECTION
Heater/defroster..................Excellent
Ventilation...........................Good
Weather sealing...................Very Good

CONSTRUCTION QUALITY
Fiberglass........................Very Good
Paint.............................Very Good
Chrome............................Very Good
Upholstery............................Good
Padding...........................Very Good
Hardware..............................Good

GENERAL
Headlight illumination............Very Good
Parking and signal lights.............Good
Wiper effectiveness...............Very Good
Service accessibility.............Very Good
Trunk space...........................Poor
Interior storage space................Fair
Bumper protection.....................Good

CHEVROLET CORVETTE 427

mortality engine, coded L88, for megalomaniacs only. The L88 also has the aluminum heads (which feature larger exhaust valves—1.84 inches vs. the standard 1.72), but with a toilet-sized single four-barrel sitting atop a hogged-out manifold. All the 427s have a suggestive hood bulge, but the L88 alone has an air intake at the rear of the hood which ducts cold air directly on the carb. The air cleaner is attached to the hood itself, and mates with the carb base via a big, spongy O-ring when the hood is closed. The L88 is all set up as if for racing—with blueprint tolerances and all the right parts—to save competitors the time and expense of tearing down a standard 427 and rebuilding it with balanced racing pistons, etc. The L88 is rated at 435 hp, like the L89, but we were almost afraid to try it. Just listening to it idle, we knew it must have over 500 real horsepower, and besides, it was Friday the 13th. Back to the safe and sane L89.

Although the Sting Ray is substantial, it doesn't feel as heavy as a 427-engined anything should. The steering is heavy to the touch, without much feel, and reasonably quick. There is enough power to break adhesion and steer with the throttle near the limit of adhesion at any speed below 100 mph, so it didn't feel nose-heavy. In fact, it felt quite neutral at, say, 65 mph, with just a little more power cranked on than necessary to hold a given radius. Among the changes for '67 was the addition of ½-inch wider wheel rims (now 6.0 inches) and wide, low-profile tires. Any car handles only as well as its tires will allow, and the Sting Ray's Akron Fats are well mated with the all-independent suspension.

The Corvette's handling is not quite matched by its ride. A Ferrari 330/GTC is harsher, but what the driver feels tells him something about the relationship between the tires and the road. The Sting Ray rides softly and vaguely—you're never sure what the car is trying to tell you. Only the Mercedes-Benz models, and to a lesser extent, the Rover 2000, the Porsche 911 and the BMW TI, have managed to combine a soft ride with a suspension system that talks to you.

The Sting Ray's four-wheel disc brakes are in a class of their own among American cars, and up to the highest standards set abroad. We have just about exhausted our cherished supply of superlatives for these brakes, so suffice it to say that they're the best, and if the Nader-Haddon axis wanted to accomplish something really useful, they should pressure Detroit into putting Sting Ray brakes on every car it builds.

Nearly all the changes for '67 are functional. The louvers behind the front wheels, for instance, really do exhaust hot engine compartment air. The optional bolt-on aluminum wheels now save weight; in the past, they had a knock-off system which made them heavier than the stock steel wheels. The pull-up handbrake between the seats is easier to get at than the old umbrella handle under the dash, but it isn't very impressive in operation. Normally, an emergency brake operates the same linings as those operated by the brake pedal, but because the Corvette's emergency brake is entirely separate (two 6.5 x 1.25-inch drums within the rear discs), the linings never get burnished—a point for owners to watch. The way to run them in is to slowly, *carefully*, pull up the handbrake—while the car is moving—every once in a while.

We have only two complaints to register about the Sting Ray. The late Ken Miles, in testing an early model Corvette, theorized "the bigger a car is outside, the smaller it is inside." This is still true; the Corvette is very large for a two-seater sports car, without a commensurate amount of room inside. We can understand why. The bucket seats aren't perfect; they lack lateral support because delicate young things have to swing their legs around to one side to get in and out. The seats are too close to the floor to gain headroom and keep a stylishly low roofline. The 427 engine does take up a lot of room that might otherwise be left in the footwells, despite the fact that the engine is offset to the right to give the driver a bit more space. The tightness around the hips, elbows and shoulders is caused by the width of the transmission tunnel. A rear-engined (or mid-engined) design would eliminate all these objections, and if Chevrolet ever builds one, it will be to improve creature comfort and decrease overall size, not to make the car handle better. But after the Corvair lawsuits, we don't think GM will ever go this route.

As it sits, the Sting Ray is the most sophisticated passenger car made in America—in terms of engine, drive train, suspension and brakes—and among the best engineered sports cars made anywhere. If that isn't good enough to make it the Best All-Around Car of 1967, we'd like to know what is. C/D

Continued from page **54**

up is blocked off in a big way in the two rear quarter areas, and the plastic rear window has the usual ripples. Draft-free ventilation is practically impossible to arrange, since there are no quarter-windows next to the "old-fashioned" wraparound windshield. The cowl vent lets in lots of air, which beats down on the shins at speed in a way that can become tiring. Fitted with a two-speed fan, the heating and defrosting system has more than enough power for this small interior.

The only major styling change for 1961 is the new tail, which most observers approved heartily. Its tucked-in lines mate well with the original shape, and even seem to increase the usable luggage volume.

Comparing the 1961 Corvette with a 1957-vintage car, one is surprised at how complete a change has been effected over the years. When the dual headlight treatment arrived, for example, the width of the hood opening was radically reduced, making work on the engine—especially spark plug changing—much more difficult. Buried down on the right side, the battery is hard to reach too. Generally, Chevy is happy with the way the fiberglass construction has worked out, one important advantage being the absolute absence of rust and the resulting high value of a used Corvette. Studies have also confirmed that integral construction (the only alternative now being considered by our industry) isn't as effective for an open car as the present frame and body layout, especially on a weight basis. So it appears that future Corvettes will stay with the present method of construction.

DETAILS AREN'T EVERYTHING

Like most General Motors cars, the Corvette boasts remarkably good detail finish, an approach that can be an asset to a sports machine. Control handles like the hand brake and hood release have handsome lettering with an impressive burnished chrome finish; other details are dealt with just as deftly. But the styling, in spite of the renovated tail, dates from an era when design effort was concentrated on the simulation of devices and effects that were mechanical in appearance but not necessarily automotive or operational: phony louvers, scoops, jet pods here and there, dashboards that sought to confuse rather than clarify.

Aerodynamics are important, certainly on a car this fast, but we realize more and more that in the automotive speed ranges there's ample room for styling within an aerodynamic area. We're glad to observe a general trend in our industry toward styling idioms that are automotive, once again, and relatively restrained in nature. The Valiant and Corvair are two excellent examples of this trend away from airplane and rocket ship orientation. Now that Chevrolet has completed its family of Corvairs, we expect the next project will be a new Corvette. With the experience they've gained with this car, and in view of the trend toward sensible styling and sensational engineering, the next Corvette should be a hum-dinger. All they have to do is put the same emphasis on *doing* things that they have in the past on *seeming* to do things. Since this 1961 Corvette already does more than most drivers can handle, that's an exciting prospect. —*SCI*

Chevrolet Corvette Sting Ray

BEST GT/SPORTS CAR OVER 3000cc
AND BEST ALL-AROUND CAR

We seem to go back and test a Corvette Sting Ray every year (there's a test in this issue), but we do it as much for the pleasure of driving the car again as we do to keep you informed of detail refinements. The Corvette was America's first sports car, and it still leads the way in American automotive engineering—the Corvette was the first American production car to feature four-wheel disc brakes. Although it isn't a big-volume seller in Chevrolet Division's line-up, the Corvette probably commands more enthusiasm and loyalty than any other GM model. Hence its third double victory in our poll, and its remarkable Best All-Around Car victory.

Continued from page **37**

the throttle a bit. On a really rough surface, the manner would be rather less unruffled, for the rear axle assembly is a heavy item of unsprung weight. After finishing the tests, we were told that the car we had been driving had not a one of the HD suspension options. We were suitably impressed.

The fact that acceleration times for this car differ somewhat from those of our previous Corvette tests is more likely due to the easy-going driving technique used than anything else. Our test driver, Mr. Rose, who was provided by GM to do the driving while the Technical Editor did the timing, confessed that standing starts were not his specialty. As we have said before, they were not the Corvette's most polished maneuver either. It is a crying shame that the new "four-link" rear suspension on the regular Chevy's is not used here, where its ability to completely eliminate axle wind-up would be most appreciated. Parenthetically, this major advance in rear suspension (for American cars, that is) comes about as an incidental result of the switch to air suspension and the attendant loss of a means of location (provided formerly by the leaf springs).

Once under way, the Positraction differential really earns its keep and the acceleration is quite breath-taking. The gear ratios in the four speed gearbox (at last!) are marvelously spaced — the ratio step between gears ranges from 1.265 to 1.325— and *all* gears are synchronized (will wonders never cease?). It is at least the equal of any gear box we've ever tried, not only with respect to the suitability of the ratios to the engine performance, but the smoothness of the synchromesh brings to mind the old metaphor about a hot knife and butter.

One fault which did show up toward the end of our acceleration runs was a trace of clutch slip when rushing the shift. When you consider that for the previous ten days this same car had been subjected to the machinations of various and sundry road-testing "experts" from all sorts of publications, then this is perhaps understandable.

Because our tests were made on a regular working day at the Proving Grounds, the normal "traffic" on the high speed straight (2½ miles of level, three-lane road in each direction with a banked turnaround at each end) prevented the Test Manager, Mr. Caswell, from allowing us to exceed 110 mph. With the same final drive ratio and engine as last year's F.I. test car, the top speed should be about the same, namely 125 mph, as the frontal aspect is not changed all that much.

As before, the throttle linkage seems a bit quicker than we would prefer, and with the faster bends requiring careful feathering, it is necessary to brace the edge of your right foot against the transmission bulge, pivoting it from there to operate the throttle. The steering wheel, in typical Chevrolet fashion, is right under the driver's chin. Even so, the Corvette is very easily controlled, the brake and clutch pedals are both well placed and smooth in operation, and there is puhlenty of room to stretch your left foot — or brace it, on sharp right turns. And brace it you must, because the Corvette's bucket-style seats are the best argument for seat belts we've seen. At the risk of repeating last year's criticisms all over again, you sit *on* them, not *in* them, and there is virtually no lateral support whatsoever. Seat belts will be standard equipment this year, which is admirable indeed; but better contoured seats would be another big step ahead, too.

The brakes were so good that we kept up our punishing test for twelve stops instead of the usual ten, and it was only in the last two that a slight but definite weakening showed up. We were therefore quite disappointed to find that these were experimental linings only. Still, it's encouraging, as it shows that Chevrolet's been doing a lot of work to provide the average Joe with significantly better brakes, without his being subjected to the drawbacks of the HD kit's Cerametallics — and with a fair amount of success.

For the price of the Corvette, check with your Chevrolet dealer; GM says they're all independent businessmen who are free to set their own prices. Especially on the options, we might add. Without quoting any figures, we'd say that on the basis of local (N.Y.) prices the Corvette ranks as a Best Buy, both as a boulevard sports car and as a competition model.

Stephen F. Wilder

Continued from page **47**

Where the original car had trouble, and where it was most painfully obvious in the Stingray, was in the brake department. The car has a rather weird braking system incorporating a servo assist and a front/rear proportioning device. In theory it works real fine but the big problem at Marlboro was the fact that Dick Thompson just didn't have enough time to get used to the thing. The twisty circuit didn't help much either. To support this we might quote Thompson after the qualifying run . . . "It's great . . . and a little more than I can handle right now. The brakes are new and a bit uneven. We'll have plenty to do in the next ten days." (The next outing would be at Danville on May 2nd.) One of the mechanics was heard to say, "It's fine dry, but if it rains tomorrow for the race, we'll finish last."

And sure enough, race day dawned dark and cloudy with rain starting during the first event. In the pit, Stingray bore signs of change; new brake fluid had been put in, a metal airscoop had been tacked on behind the cockpit and the twin set of hood louvers had been completely cut out. All these modifications seemed to indicate an attempt to improve braking.

Continuous light showers made the track even more treacherous and almost every lap saw a spinout, most frequently at Cappy's Corner . . . a tight loop that gives trouble to small production cars, not to mention the big C Modified machines. By the time the feature event was to be run the stretch leading into the corner had sent many a production Corvette fishtailing while braking for the turn.

On the grid the Stingray lined up for its first race beside an RSK Porsche, a Lister Corvette and a Lister Jag. With only fifteen minutes logged in this machine, Thompson faced not only tough and tested opposition, but a slick and twisting track ill-suited for a car with the Stingray's handling characteristics. Once off the grid, the car stayed with the other front line starters but seemed to have trouble controlling wheelspin.

The Stingray made its best showing on the ¾ mile back straight where it flew past the now lapped stragglers. It was harder for Thompson to get through the corners which were now twice as difficult with traffic. At midrace Cappy's Corner claimed the Stingray as Thompson spun out and lost a half lap to the leaders. Returning to the track Dick had his hands full trying to catch up, trying various front/rear brake combinations in an effort to make up time. During the last quarter of the race he became aware of overheating problems and began to stroke it in order to prevent possible damage and to get better acquainted with the brakes. When the rear ones got troublesome he shut them off for four laps then switched on again to find that the brief rest had perked them up. Even with all the trouble in the corners the Stingray gobbled up the straight stretches to finish fourth behind two RSKs and the Lister Jag.

Obviously pleased by the results, Mitchell stated that, "The Stingray is no longer a show car. Now we're ready to race," but he refused to comment on future plans or possible European races.

Judging from its brief but promising performance there's going to be a lot of action when some of the handling difficulties are worked out, and it's quite logical to assume that the Stingray will be whetting its barbed tail with a diet of dung-beetles in mind.

—du